SpringerBriefs in Energy

For further volumes:
http://www.springer.com/series/8903

Michael D. Max · Arthur H. Johnson
William P. Dillon

Natural Gas Hydrate - Arctic Ocean Deepwater Resource Potential

Michael D. Max
Arthur H. Johnson
William P. Dillon
Hydrate Energy International, Inc.
Kenner, LA
USA

ISSN 2191-5520 　　　　　　ISSN 2191-5539 (electronic)
ISBN 978-3-319-02507-0　　　ISBN 978-3-319-02508-7 (eBook)
DOI 10.1007/978-3-319-02508-7
Springer Cham Heidelberg New York Dordrecht London

Library of Congress Control Number: 2013949606

© The Author(s) 2013
This work is subject to copyright. All rights are reserved by the Publisher, whether the whole or part of the material is concerned, specifically the rights of translation, reprinting, reuse of illustrations, recitation, broadcasting, reproduction on microfilms or in any other physical way, and transmission or information storage and retrieval, electronic adaptation, computer software, or by similar or dissimilar methodology now known or hereafter developed. Exempted from this legal reservation are brief excerpts in connection with reviews or scholarly analysis or material supplied specifically for the purpose of being entered and executed on a computer system, for exclusive use by the purchaser of the work. Duplication of this publication or parts thereof is permitted only under the provisions of the Copyright Law of the Publisher's location, in its current version, and permission for use must always be obtained from Springer. Permissions for use may be obtained through Rights Link at the Copyright Clearance Center. Violations are liable to prosecution under the respective Copyright Law.
The use of general descriptive names, registered names, trademarks, service marks, etc. in this publication does not imply, even in the absence of a specific statement, that such names are exempt from the relevant protective laws and regulations and therefore free for general use.
While the advice and information in this book are believed to be true and accurate at the date of publication, neither the authors nor the editors nor the publisher can accept any legal responsibility for any errors or omissions that may be made. The publisher makes no warranty, express or implied, with respect to the material contained herein.

Printed on acid-free paper

Springer is part of Springer Science+Business Media (www.springer.com)

Preface

Natural gas hydrate (NGH) is the last of the recognized unconventional resources of natural gas and perhaps the greatest. Coalbed methane, for which depressurization proved to be the critical factor, and tight/shale gas, for which fracking and lateral, long-pay section drilling are key, have radically altered the indigenous gas resource/reserves in the United States. A median global resource potential for high grade NGH sands, which are deepwater host sediments for NGH based on a new petroleum system approach, may have as much as 43,300 trillion cubic feet (TCF) gas-in-place, of which 50 % may be technically recoverable (Johnson 2011). This compares with resource and reserve estimates for coalbed methane of 9,000 TCF, shale gas of 16,000 TCF, and tight gas of 7,400 TCF (NPC 2007).

There is increasing evidence that natural gas can be produced from high-grade NGH concentrations (Max et al. 2006) hosted by coarse-grained and sandy sediments using existing conventional oil and gas production technology (Moridis and Kowalsky 2006). In addition, the physical nature and location of NGH may allow much less expensive innovative technology to be used in drilling and other aspects of exploration and production. While the volume of natural gas (NG) contained in the world's NGH accumulations may greatly exceed that of other NG resources (Collett 2002), a substantial proportion of NGH is present in low-grade accumulations (Boswell and Collett 2011) that are unlikely to be developed commercially (Moridis and Sloan 2007) using existing technology and methods. Innovation and new production techniques may also render the natural gas in these low grade NGH deposits commercial, which would have the effect of greatly enlarging the NGH resource base.

Many have dismissed oceanic NGH as a gas resource for the far future. However, the world's first technical production test of oceanic NGH was carried out on the 40 TCF Nankai NGH deposit according to a planned timeline (Kurihara et al. 2011) during March 2013 by JOGMEC (2013). The deposit is scheduled for production in 2018, which is only 5 years from the first production test. This is a development timeline consistent with conventional deepwater field development. Commercial production of NGH off Japan is likely because natural gas produced from the Nankai NGH deposit should compete well with the rather high delivered price of liquefied natural gas (LNG) that has been in the $15–$18 MMcf range in the 2011–2013 time period. We submit that once the Japanese begin to be able to replace gas imports with indigenous production from NGH, this will both affect the world LNG business and stimulate other NGH development. In addition to the Japanese, Korea, India, and China have aggressive NGH exploration programs in which drilling and production tests are scheduled (as of late summer 2013) for 2013 and 2014.

Recent development of shale gas in North America has provided a large gas supply that may delay the development of NGH there. In other countries that have NGH potential, however, development of indigenous gas supply for energy security is a national priority. Although NGH will likely be developed by countries for which NGH is an indigenous potential resource, its commercialization is in its infancy. As development takes place, and new and improved technologies are brought into play, it is possible that NGH may become competitive with other gas resources on a produced price basis. We suggest that commercialization of NGH would have the same disruptive effect on the world gas market that the development of land-based unconventional resources in North America has achieved since 2005.

The Arctic Ocean provides a particularly useful area in which to apply the general principles of assessing NGH resource potential from the different aspects of the NGH petroleum system. The Arctic Ocean is compact, has a related geological history for potential host sediments and is a recognized hydrocarbon province, and is affected broadly by the same weather and climate conditions. The principles applied here can be applied along any other continental margin, particularly those that have Pleistocene glacial history, in order to determine NGH potential.

References

Boswell R, Collett TS (2011) Current perspectives on gas hydrate resources. Energy Environ Sci 4:1206–1215. doi:10.1039/c0ee00203h
Collett, T. S. 2002. Energy resource potential of natural gas hydrates: American Association of Petroleum Geologists Bulletin 86, 1971–1992.
JOGMEC (2013) News release. Gas production from methane hydrate layers confirmed. www.jogmec.go.jp. Accessed 12 March 2013, p 3
Johnson AH (2011) Global resource potential of gas hydrate—a new calculation. Fire in the ice. NETL, U.S. Department of Energy 11(2):1–4
Kurihara M, Ouchi H, Sato A, Yamamoto K, Noguchi S, Narita J, Nagao N, Masuda Y (2011) Prediction of performance of methane hydrate production tests in the eastern Nankai Trough. Proceedings of the 7th international conference on gas hydrates (ICGH 2011), Edinburgh, Scotland, United Kingdom, 17–21 July 2011, p 16
Max MD, Johnson A , Dillon WP (2006) Economic geology of natural gas hydrate. Springer, Berlin, Dordrecht, p 341
Moridis GJ, Kowalsky M (2006) Gas production from unconfined Class 2 oceanic hydrate accumulations. In: Max MD (ed) Natural gas hydrate: in oceanic and permafrost environments, 2nd edn. Kluwer Academic Publishers (now Springer), London, Boston, Dordrecht, p 249–266
Moridis, G.J. & Sloan, E.D. 2007. Gas production potential of disperse low-saturation hydrateaccoumulations in oceanic sediments. Energy Conversion and Management 48, 1834-1849.
NPC. 2007. Topic Paper #29, Unconventional Gas. Working Document of the NPC Global Oil& Gas Study. Team Leader, Holditch, S.A. et al. National Petroleum Council, 52pp.<www.npc.org/study_topic_papers/29-ttg-unconventional-gas.pdf>.

Acknowledgments

Thanks to Tim Collett (U.S. Geological Survey) and Ray Boswell (U.S. Department of Energy), principal among the many colleagues who have brought the unconventional NGH resource from a scientific curiosity to the threshold of responsible commercial development. Special thanks also to Jurgen Meinert (Institute for Geology, University of Tromsø, Norway).

Contents

1 **The Arctic Ocean**... 1
   1.1   Tectonics of the Arctic Basin 3
   References 6

2 **Sediment Delivery Systems; Ice, Rivers and the Continental Margin**... 9
   2.1   East Alaska—North American Arctic Islands—Greenland Margin 12
   2.2   Greenland 12
   2.3   Barents—Kara Seas Margin............................... 13
        2.3.1   West Barents Margin 13
        2.3.2   NW Barents Margin 14
        2.3.3   St. Anna Trough 14
        2.3.4   Kara Sea—Eastern Margin....................... 14
   2.4   Laptev Sea—West Siberian Arctic Sea Zone................... 15
   2.5   East Siberian Sea Zone.................................... 16
   References 17

3 **Natural Gas Hydrate: Environmentally Responsive Sequestration of Natural Gas**.. 19
   References 23

4 **NGH as an Unconventional Energy Resource** 25
   4.1   Permafrost NGH.. 26
        4.1.1   Geologically Trapped............................... 27
        4.1.2   Permafrost-Related, Non-Geologically Trapped 28
        4.1.3   Vein-Type Deposits 28
   4.2   Oceanic NGH.. 29
   References 30

5	**Elements of the NGH Petroleum System**		33
	5.1	Sufficient Gas Source/Flux and the BSR	35
	5.2	Migration Pathways/Feeding the Thermodynamic Trap in the GHSZ	38
	5.3	NGH High-Grade Reservoirs	40
		5.3.1 GHSZ Thickness	41
		5.3.2 Suitable Sediment Hosts (Turbidite Sands)	41
	References		44
6	**Path to NGH Commercialization**		47
	References		52
7	**Gas Production from NGH: We Have All the Basic Tools**		55
	7.1	Phase 1. Basin Analysis	57
	7.2	Phase 2. Potential Reservoir Localization	58
	7.3	Phase 3. Deposit Characterization and Valuation	58
	References		59
8	**What More Do We Need to Know?**		61
	8.1	Exploration Factors	62
		8.1.1 BSR Identification from Reflection Seismic Data	63
		8.1.2 Heat Flow Data/Geothermal Gradients	63
		8.1.3 Natural Gas Migration Path Analysis	65
		8.1.4 Exploration Drilling	65
	8.2	Production Factors	66
		8.2.1 Drilling in Preparation for Gas Production	66
		8.2.2 Thermodynamic Models for NGH Conversion	66
		8.2.3 Geotechnical Models	67
		8.2.4 Flow Assurance	68
		8.2.5 Logistics and Infrastructure	68
	8.3	NGH-Specific Technology Opportunities	69
		8.3.1 Moving to the Seafloor	71
		8.3.2 Drilling and Logging	71
		8.3.3 Undersea Processing and Completions	72
		8.3.4 NGH: Specific Vessels/Seismic Survey	72
	8.4	Will NGH Deposits Be Commercially Competitive with Conventional Gas?	73
	References		74
9	**NGH Likelihood in the Arctic Ocean**		77
	References		83

10	**Estimates of the NGH Resource Base in the Arctic Region**.........	85
	References...	88
11	**Oceanic NGH: Low Risk Resource in Fragile Arctic Environment**	91
	11.1 Risk Factors of Conventional Hydrocarbon Production	93
	11.2 Inherent Geosafety of NGH Production.....................	93
	References...	97
12	**Economic and Political Factors Bearing on NGH Commercialization**	99
	References...	101
13	**Logistical Factors for Arctic NGH Commercialization**............	103
	References...	106
14	**Natural Gas as Fuel and Renewable Energy Aspects**	109
	References...	110
Index..		111

Chapter 1
The Arctic Ocean

Abstract The Arctic Ocean is an almost entirely enclosed basin floored by two major tectonic zones, the Eurasia Basin that flanks the European continental margin, which is floored by a relatively regularly disposed oceanic crust at abyssal depths, and the Amerasia Basin, which has an older and more complex tectonic history. Broad continental shelves along the European and Asian continental margins contrast with much narrower continental shelves along the North American and Greenland continental margins. The continental margins of the Arctic Ocean are generally draped with sediments derived as a result of the harsh weathering and erosional framework of the Pleistocene glaciations. The factors that have the strongest effect on the degree of sediment winnowing and composition in the slope depositional environment are the width of the shelf and the degree of ice cover on the continental shelf segment and the edge of the ice to the continental slope break. The clastic sediments, which can be expected to comprise a high percentage of the continental slope and deep continental shelf turbidites, are the focus for natural gas hydrate concentrations.

Keywords Continental slope • Continental margin • Turbidites • Sands • Clastic sediments • Sediment host • Natural gas hydrate • NGH

The Arctic Ocean is almost entirely enclosed basin that comprises less than 3 % of the area of the World Ocean. Very broad continental shelves along the west Alaskan and Asian and European continents from the Chukchi to the Barents Sea flank the abyssal region, whereas the North American and Greenland continental shelves are narrow. The shallow-water Bering shelf between Alaska and easternmost Russia (Fig. 1.1) only allows exchange of surface waters. The deep-water passage into the North Atlantic between Svalbard and Greenland, however, allows deep, cold, highly saline water to flood into the world ocean at depth. Surface water inflows on and along the Barents Sea shelf and through Bering Strait are entrained by anti-clockwise currents (Jones 2001), with the only significant shallow

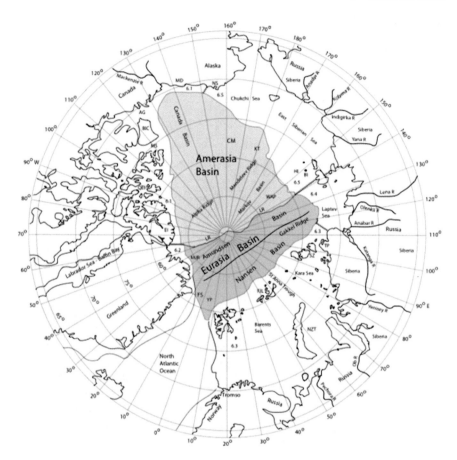

Fig. 1.1 Polar projection location map after Jakobsson et al. (2004, 2008). *AG* Amundsen Gulf, *BI* Bennett Island, *BIC* Banks Island, *CM* Chukchi Microcontinent, *EI* Ellesmere Island, *FS* Fram Strait, *FJL* Franz Joseph Land, *HI* Henrietta Island, *LR* Lomonosov Ridge, *MD* Mackenzie Delta, *MJR* Morris Jessup Rise, *MS* M'Clure Strait, *NS* North Slope, *NZT* Novaya Zemyla Trough, *QEI* Queen Elizabeth Islands, *SZ* Severnaya Zemlya, *SV* Svalbard Archipelago, *TP* Taymyrskiy Peninsula, *YP* Yermak Plateau. Heavy *dashed lines* connecting coastlines with continental shelf edges are boundaries between the tectono-sedimentary areas: 6.1., east Alaska–north American Arctic islands (older sediment, glaciated except in W), 6.2 Greenland zone (older sediments, glaciated), 6.3 Barents–Kara seas zone (younger sediments, glaciated), 6.4. Laptev sea–west Siberian sea zone (younger sediments, glaciated), 6.5. East Siberian–Chukchi sea zone (younger sediments, unglaciated except in W)

water outflow along the east coast of Greenland. This deep water exits the Arctic Ocean Basin into the North Atlantic through the abyssal Fram Strait as the Transpolar Current. This deep plunging water from both Polar Regions is responsible for the characteristic lowering of temperature with increasing depth in the world ocean.

We are concerned here only with those geological and other physical elements critical to the formation of potentially commercial deposits of oceanic NGH.

According to NGH petroleum system analysis, which will be described in greater detail subsequently, these deposits will be restricted to deep continental shelf and continental slope regions. Therefore, we only consider the geology of the continental shelf areas from the point of view of their contribution of sediment to deep trenches in the continental shelves and the flanks of the abyssal Arctic Basin. We do not discuss conventional hydrocarbons and perhaps relict permafrost hydrate or gas venting from continental shelf regions in those shelf areas that were unglaciated in the Arctic continental shelves.

Although concerned by the recent warming trends in the Arctic, we do not discuss potential relationships between methane and its potential contribution to the atmospheric greenhouse. We do comment, however, on some of the aspects of Arctic warming, which will have a considerable impact upon increased hydrocarbon exploration and production, as well as shipping traffic and the potential for human-induced pollution in the Arctic Ocean region.

1.1 Tectonics of the Arctic Basin

The Arctic Basin consists of two major abyssal basins (Fig. 1.1), the Amerasia Basin, adjacent to the North American and Asian landmasses and the Eurasia Basin, occupying the European-fronting sea area, separated by the Lomonosov Ridge. In general, and as pointed out by Miller et al. (2008) and Miller and Verzhbitsky (2009), who focus on the genesis of the Amerasia Basin north of the East Siberian Shelf, the ice cover of the Arctic Ocean conceals a number of the few remaining unresolved plate tectonic puzzles on Earth. The geological and tectonic history of the Amerasia Basin is particularly obscured, due to its having greater year-round ice cover. It also lacks well-defined magnetic or gravity patterns that could suggest oceanic crust generation and orientation.

The two basins are described separately because the tectonics controlling their generation and the sediments deposited on the continental margins of the basins are different, especially in their older histories. We use the term 'continental margin' to mean approximately the continental slope sedimentary deposits that may host NGH deposits that rest on the steeply sloping margin of upstanding continental crust. This is a tectono-sedimentary use of the term rather than referring to all continental crust areas seaward of present coastlines. Around the Arctic Basin, this is a dramatic geomorphologic feature that forms the margin to both the Amerasia and Eurasia Basins. Thicker and older sedimentary successions have a greater likelihood to host more gas source beds. The younger sedimentary successions, which are common in many ways, differ primarily depending on whether the continental shelves facing the basins were covered by ice caps or were only permafrost regimes.

The Amerasia Basin contain an older, complex Mesozoic history of crustal thinning, and rifting in an extensional plate framework along with the probable development of restricted oceanic crust segments and microplates (Grantz et al. 2011). Herron et al. (1974) suggested that the Amerasia Basin is floored by oceanic crust as old as the Jurassic magnetic quiet period from 180 to 150 Ma, which is reflected in modern seafloor maps (CIT 2009). Miller and Verzhbitsky (2009), suggest that Late Jurassic-Early Cretaceous sedimentary rocks as young as c. 140–135 Ma are involved in the last stages of crustal shortening that began in this region in the Late Jurassic.

The relatively few heat flow measurements in the deeper parts of the Arctic Ocean Basin suggest relatively low heat flow (Hasterok and Chapman 2008; Hasterok et al. 2011), supporting the concept of oceanic crust as old as Jurassic. But until scientific drilling in the most controversial parts of the Arctic Ocean seafloor can take place, many questions about the origin of the Amerasia Basin will remain unresolved. Tectonic activity associated with formation of the Amerasia Basin appears to have ceased by the time volcanism began in the Siberian Okhotsk-region at about 90 Ma (Miller and Verzhbitsky 2009).

A considerable history of geological events predated the breakthrough of the northern propagation of the Atlantic Ocean constructive plate margin and the formation of the Eurasia Basin. The Canada Basin segment at the western North American end of the Amerasia Basin (Fig. 1.1) has a relatively smooth abyssal seafloor, suggesting the presence of underlying oceanic crust, as does the Wrangel Abyssal Plain off the East Siberian Sea Shelf and a small abyssal plain to north of the central Lomonosov Ridge. The Siberian flank of the Amerasia basin, between the Canada and Wrangle Basins is structurally more complex, with the Chukchi microcontinent (Grantz et al. 2011) and the Siberian end of the Alpha-Mendeleev Ridges and basin areas in the northern Amerasia Basin that is referred to as the Large Igneous Province by Grantz et al. (2011). Structural and plate tectonic relationships are unclear in the region. There are probably a number of small spreading centers, possibly thick igneous and volcanic rocks, and thinned continental fragments not dissimilar in their geological context from areas of the Mediterranean Seafloor.

It would appear that the general tectonic division of the Amerasia Basin into a northern segment, containing a complex floor of oceanic crust, thinned continental fragments and the Alpha-Mendeleev Large Igneous Province, and the southern Canada Basin that is mainly underlain by relatively straightforward oceanic crust is consistent with morphology (Miller et al. 2008). Opening of the Amerasia Basin was allowed by transform and/or shear along the southern margin of the Lomonosov Ridge, which suffered tectonic subsidence associated with its separation from the European crust at about 56 Ma (Minakov and Podiadchikov 2012). A conjugate shear related to this line of shear structures may have been rejuvenated as the nexus of the boundary between continental crust sliver of the Lomonosov fragment and the oceanic crust of the Eurasia Basin.

As noted by Max and Lowrie (1993), the geological history of the northern Gulf of Mexico (GoM) and sedimentation in the Amerasia Basin have analogous geological histories. This has implications for NGH in the Arctic as new NGH information from the GoM offers encouraging perspectives. The northern GoM is also underlain by Later Jurassic oceanic crust of about 166–150 Ma (Bird et al. 2005) and thus will floor overlying sediments of comparable age.

The precise history regarding the orientation of spreading in both the northern GoM and the Amerasia Basin, or possible rotation of marginal continental masses, is unresolved.

We are less concerned with the oldest tectonic history than we are with the more recent development of the existing passive and transform continental-oceanic margins along which potential NGH sediment hosts are draped. We follow Grantz et al. (2011) in regarding all the present margins of the Amerasia Basin as having initially formed as either fundamentally extensional or strike-slip margins, except for the eastern margin of the Chukchi microcontinent that may have a compressional component related to part of its rotational hinge against the Chukchi margin or crust within the basin. A more detailed consideration for locations of those areas in the continental margin where basins containing sediments that could be methane producing could provide deeper sourced methane into the overlying GHSZ is left to a more detailed workup of economic potential.

The Cenozoic Eurasia Basin adjacent to Europe and western Russia (Fig. 1.1) is the northernmost prolongation of the actively spreading Atlantic Ocean constructive plate margin system. Morphologically and structurally well-defined margins flank the axial constructive plate margin marked by the Gakkel Ridge. This separates the basin into two elongate abyssal regions, with the southern margin of the Nansen Basin flanking the northern Barents Sea Shelf having thicker sediments as a result of continued sedimentation from the continent. Urlaub et al. (2010) demonstrate an almost symmetrical ocean crust beneath the paired Basins (Fig. 1.1) for this slowest spreading segment of oceanic crust. A transform or sheared margin along the Laptev Sea continental-oceanic crust junction terminates the almost rectilinear Eurasia Basin.

The SW margin near Greenland may have once been impinged upon by Greenland during the opening of the Labrador Sea, but no large-scale distortion of the spreading center along the Gakkel Ridge appears to have taken place, and Greenland is now passive with respect to the North American plate (Brozena et al. 2003). The Yermak Plateau, which appears to be prolongations of abandoned and thinned continental crust prongs with possibly some basic igneous/volcanic intrusive rocks related to the opening of the Eurasia Basin (Jokat et al. 2008) and the Morris Jessup Rise north of Greenland break the otherwise rectilinear symmetry of the Atlantic end of the Eurasia Basin. In contrast to the unresolved complexity of the Amerasia Basin, the Eurasia Basin is relatively well defined by magnetic striping, with the Lomonosov Ridge to the north apparently a structurally thin, continuous fragment of the European-Siberian continental conjugate margin of the Barents and Kara Seas that was detached when the Eurasia Basin opened (Jokat et al. 1995).

The broad Eurasia continental shelf (Cherkis et al. 1991) hosts many prospective conventional hydrocarbon sedimentary basins (Larssen et al. 2005; Drachev et al. 2010) as do the narrower shelves of Alaska (Attanasi and Freeman 2009) and Canada (Drummond 2000; Drummond et al. 2000). It would appear that the entire Arctic is a petroleum province (Spenser et al. 2011). The shelves are too shallow, however, to host more than traces of NGH in thin, impersistent GHSZs except in the deeper troughs (Wood and Jung 2008; Long et al. 2008). Thus, it is only in the deeper water the continental margin sediments, and possibly in subsided continental fragments, that significant concentrations of NGH may occur.

The continental margins of the Amerasia and Eurasia Basins are the immediate focus of oceanic NGH exploration because gas generation and the GHSZ coincide in continental margin sediments. Deepwater sands are emerging as the primary host for NGH concentrations on continental slopes, where the GHSZ is thick enough to provide for significant development. The main importance of the shelves is their relationship to sedimentation on the adjacent continental slopes, and the degree to which they could have acted as effective filters for sand emanating from subaerial erosion during high stands, and as conduits for those sands to reach the continental slopes during lowstands. In general, broad continental shelves may have a greater potential to sequester coarser grained sediments, especially during low stands when deposition in river valleys and deltas on the gently sloping shelf regions will be common. However, if drainage systems have definite, long standing channelization, considerable sands can reach the continental slope in depths where NGH formation and concentration may be significant. Regions with narrow continental shelves (Fig. 1.1) are much weaker barriers to sand deposition on the continental slopes.

References

Attanasi ED, Freeman PA (2009) Economics of undiscovered oil and gas in the North slope of Alaska: economic update and synthesis. U.S. Department of the Interior U.S. Geological Survey Open-File Report 2009–1112, p 65

Bird DE, Burke K, Hall SA, Casey JF (2005) Gulf of Mexico tectonic history: hotspot tracks, crustal boundaries, and early salt distribution. Am Assoc Pet Geol 89(3):311–323

Brozena JM, Childers VA, Lawver LA, Forsberg R, Faleide JI, Eldholm O (2003) New aerogeophysical study of the Eurasia Basin and Lomonosov Ridge: Implications for basin development. Geology 31(9):825–828

Cherkis NZ, Fleming HS, Max MD, Czarnecki MF (with Kristoffersen Y, Midthassel A, Roboengen K) (1991) Bathymetry of the Barents and Kara Seas. Scale: 1:2,313,000. Geological Society of America Map and Chart Series MCH047 (1 sheet)

CIT (2009) Seafloor age. California Institute of Technology. Referenced from Muller RD, Sdrolias M, Gaina C, Roest WR 2008. Age, spreading rates and spreading symmetry of the world's ocean crust. Geochem Geophys Geosyst 9:Q040060. doi:10.1029/2007GC001743

Drachev SS, Malyshev NA, Nikishin AM (2010) Tectonic history and petroleum geology of the Russian Arctic Shelves: an overview. Geological society, London, petroleum geology conference series 2010, vol 7, pp 591–619. doi:10.1144/!0!0070591

Drummond KJ (2000) The energy resources map of the arctic sheet of the circum- pacific region is a compilation at a scale of 1:10,000,000. USGS map CP-51

Drummond KJ, Moore GW, Swint-Iki TR (2000) Explanatory notes for the energy-resources map of the circum-pacific region, arctic sheet CP-51. U.S. geological survey, p 30

Grantz A, Hart PE, Childers VA (2011) Development of the Amerasia and Canadian Basins, Arctic Ocean. In: Spenser AM, Embry AF, Gautier DL, Stompkova AV, Sorensen K (eds) Arctic petroleum geology. Geological Society of London Memoir, vol 35, pp 771–799. doi:10.1144/M35.50

References

Hasterok D, Chapman DS (2008) Global heat flow: a new database, a new approach. EOS Trans Am Geophys Union. Fall Meeting Supplement 89, T21C-1985

Hasterok D, Chapman DS, Davis EE (2011) Oceanic heat flow: implications for global heat loss. Earth Planet Sci Lett 311:386–395. doi:10.1016/j.epsl.2011.09.044

Herron EM, Dewey JF, Pitman WC (1974) Plate tectonic model for the evolution of the Arctic. Geology 2(8):377–380. doi:10.1130/0091-7613(1974)2<377:PMFTE>2.0.CO;2

Jakobsson M, Mcnab R, Cherkis N, Shenke H-W (2004) The international map of the Arctic ocean (IBCAO). Polar stereographic projection, scale 1:6,000,000. Research publication RP-2. U.S. National Physical Data Center, Boulder, Colorado 90305

Jakobsson M, Macnab R, Mayer L, Anderson R, Edwards M, Hatzky J, Schenke H-W, Johnson P (2008) An improved bathymetric portrayal of the Arctic ocean: implications for ocean modeling and geological, geophysical and oceanographic analyses. Geophys Res Lett 35(5):L07602. doi:10.1029/2008GL033520

Jokat W, Weigelt E, Kristophersen E, Rasmussen Y, Schone T (1995) New insights into the evolution of the Lomonosov Ridge and the Eurasia Basin. Geophys J Int 122:378–392

Jokat W, Geissler W, Voss M (2008) Basement structure of the north-western Yermak Plateau. Geophys Res Lett 35(L05309):6. doi:10.1029/2007GL032892

Jones EP (2001) Circulation in the Arctic ocean. Polar Res 20(2):139–146

Larssen GB, Elvebakk G, Henriksen LB, Kristensen S-E, Nilsson I, Samuelsberg TJ, Svana TA Stemmerick L, Worsley D (2005) Upper Palaeozoic lithostratigraphy of the southern part of the Norwegian Barents Sea. NGU Norges geologiske undersøkelse Geolological Survey of Norway Bulletin 444, p 45

Long PE, Wurstner SK, Sullivan EC, Schaef HT, Bradley DJ (2008) Preliminary geospatial analysis of arctic ocean hydrocarbon resources. U.S. department of energy/pacific northwest national laboratory PNNL-17922

Max MD, Lowrie A (1993) Natural gas hydrates: Arctic and Nordic Sea potential. In: Vorren TO, Bergsager E, Dahl-Stamnes ØA, Holter E, Johansen B, Lie E, Lund TB (eds) Arctic geology and petroleum potential. Proceedings of the Norwegian petroleum society conference, 15–17 August 1990. Tromsø, Norway. Norwegian Petroleum Society (NPF), Special Publication 2 Elsevier, Amsterdam, pp 27–53

Miller EL, Verzhbitsky V (2009) Structural studies near Pevek Russia: implications for formation of the east Siberian shelf and Makarov basin of the Arctic ocean. In: Stone DB et al. (eds) Geology, geophysics and tectonics of northeastern Russia: a tribute to L. Parfenov, Stephan Mueller special publication series 8, European Geophysical Union, pp 223–241

Miller EL, Soloviev A, Kuzmichev A, Gehrels G, Toro J, Tuchkova M (2008) Jura-Cretaceous syn-orogenic deposits of the Russian Arctic: separated by birth of Makarov basin? Norw J Geol 88(4):201–226

Minakov AN, Podiadchikov YY (2012) Tectonic subsidence of the Lomonosov Ridge. Geology 40:99–102. doi:10.11130/G32445.1

Spenser AM, Embry AF, Gautier DL, Stoupakova A, Sorensen K (2011) Arctic petroleum geology. Geological society of London memoir M0035. ISBN: 978-1-86239-328-8, p 818

Urlaub M, Schmidt-Aursch MC, Jokat W, Kaul N (2010) Gravity crustal models and heat flow measurements for the Eurasia Basin, Arctic Ocean. Mar Geophys Res 30:277–292. doi:10.1007/s11001-010-9093-x

Wood WT, Jung W-Y (2008) Modeling the extent of earth's marine methane hydrate cryosphere. Proceedings of the 6th international conference on gas hydrates (ICGH 2008), Vancouver, British Columbia, Canada, 6–10 Jul 2008

Chapter 2
Sediment Delivery Systems; Ice, Rivers and the Continental Margin

Abstract Pleistocene glacial sediments will predominantly host NGH, and possibly only those to depths of no more than 1 km. Older sediments will likely be buried too deeply to host NGH. Each glacial episode would have produced a suite of sediments related to the onset and glacial maximum period and especially during the onset of the following interglacial period when the melting of the ice cap would produce large volumes of water that would strongly affect sediment winnowing and transport. These periods of maximum water flow would be likely to produce the clastic sandy sediments that would be ideal hosts of high-grade NGH deposits in the deeper continental shelves and the continental slopes. In addition, sea level variation would have strongly controlled the position of the shoreline positions during the glacial and interglacial cycles. Sediments within about 1.2 km depth below the seafloor comprise the exploration zone for NGH and related gas deposits. The continental margins of the Arctic Ocean have been divided into 5 regions for analysis of the degree to which they could provide optimal NGH host sediments to suitable depositional environments.

Keywords Barents • Laptev • East Siberian • Alaska • Canada • Greenland • Sediment host • Natural gas hydrate • NGH

The tectono-sedimentary framework of those sediments found within the GHSZ that could act as hosts for NGH concentration and for methane generation below the GHS is fundamental to assessing NGH potential. The continental shelves of the Arctic Ocean have been subjected to the periodic glacial and interglacial episodes of the Pleistocene. Some continental shelf areas were covered with icecap while other areas were only permafrost regions. This factor, and the width of the continental shelves that strongly affect the degree of sediment winnowing and transportation are the basic controls of the types of sediment found in the deeper parts of the continental shelves and on the slopes. In configuring sediment models that are important for localization of NGH, only the Pleistocene glacial episodes, and possibly only the most recent ones need to be understood in detail because older sediments will likely be buried too deeply to host NGH.

Each glacial episode would have produced a changing suite of sediments related to glacial erosion during the onset and glacial maximum period and especially during the onset of the interglacial when the melting of ice cap would produce large volumes of water flow that would affect sediment winnowing and transport. In addition, sea level varies considerably during glacial cycles. In the Arctic Ocean, large areas of continental shelf, particularly in the East Siberian Sea (Fig. 1.1) would have been exposed and the continental shelves would have been as narrow as they are presently along the North American Arctic Ocean continental margin.

Because of the periodic sediment generation and transport, those continental margin sediments that would have drained from ice cap areas into abyssal depths can be expected to contain more sandy and coarse grained sediment that provide an optimal host for NGH. In addition to the sediments on the continental slope proper, which may be dominantly turbidite in nature, sediments in the deep trough areas and in older sediments in the deeper continental margins may also host NGH. Our estimates of NGH are based only on those sediments in the continental slope sedimentary wedge and deeper parts of continental shelves, because that is where the known analogues exist from which we extrapolate. For instance, in addition to the recent sedimentological model proved in the Walker Ridge drilling (Frye et al. 2011), Egawa et al. (2013) have shown that sandy turbidite accumulation in a hinged basin provided the framework for the deposition of the NGH host sand sediments of the Nankai deposit.

Slope depositional regions for the Arctic Ocean Basin can be defined based on the width, water depth, and general morphology of the shelves, their anticipated sediment load and its winnowing and likelihood of reaching a final repository on the continental margin and in the basins. The continental masses and their often-emergent continental shelves have been repeatedly subjected to glacial and interglacial episodes that have generated considerable sediment, much of it probably coarse grained. In addition, the basins and their margins are not the same age. Because the sediments on the continental margins are a product of both the tectono-sedimentary framework (i.e., oldest geological age of the basins and their margins), and the depositional transport system (ice streams, fluvial, permafrost), the region has been divided into five zones (Table 2.1, Fig. 2.1) in order to identify those regions where NGH host sediments would be most likely to be found in quantity.

Table 2.1 Summary of continental margin major basin age and sediment depositional systems

Zone	Amerasia Basin (older)	Eurasia Basin (younger)
6.1. E. Alaska-N. America	Ice streams—fluvial, esp/ in mackenzie margin	
6.2. Arctic greenland	Ice streams—fluvial	
6.3. Barents-Kara sea		Ice streams—fluvial
6.4. Laptev sea		Ice edge permafrost mixed—fluvial
6.5. E. Siberian-Chukchi seas	Permafrost—fluvial	

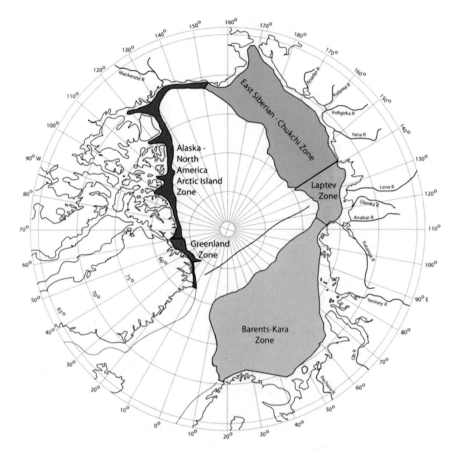

Fig. 2.1 Continental shelf sediment delivery zones for continental margin sediments of the Arctic Basin. Polar projection location map after Jakobsson et al. (2004, 2008)

The Amerasia Basin offers the greater likelihood of hosting commercial NGH deposits because it contains the most widespread thick sediments Max and Lowrie (1993). All this sediment has been delivered by the north-flowing river systems in North America and Asia (Siberian shelf). Sediments in the younger Eurasia Basin (Fig. 2.1) are neither as thick nor as widespread but may be thick enough to host significant NGH concentrations. However, greater sediment thickness enhances the potential of the sediments to generate natural gas and to host NGH concentrations. Sediments in the Eurasia Basin are thinner than those in the Amerasia Basin and the GHSZ is also thinner because of the higher geothermal gradients.

Because the composition of the sediments is liable to be biased toward sands, there is good potential for NGH deposits in the Eurasia Basin also. The tectono-sedimentary zones that are based on the age of the continental margins have a further major control on sedimentation in the ocean basins. Some of the continental margins of the ocean basins were ice-covered while others were subject to permafrost conditions (Table 2.1). These factors largely determine the amount and type of sediment both on the continental shelf and along a segment of continental margin.

2.1 East Alaska—North American Arctic Islands—Greenland Margin

This zone has a thickly sedimented continental margin that, at its base, locally rests on older oceanic crust of the Amerasia Basin. This marginal zone reaches eastward from offshore Alaska to about the northern tip of Greenland. The western boundary with the East Siberian Sea is picked about where the northern base of the coalesced deepwater sediment fans from the Mackenzie River abut the northern Alaska shelf. Although this thick sediment pile (Max and Lowrie 1993) has not been derived directly from an incised, glaciated margin as has that to the east along the Canadian Islands, the sediment appears to be generally derived from the western flank of the Canadian icecap. The margin along NE Greenland has relatively little marine sediment on its continental slope but is included in this zone for simplicity. Although the Alaskan portion of this zone now has virtually no large rivers and possibly relatively light Plio-Pleistocene sedimentation, the thickest marine sediment accumulation in the Arctic is banked against this American—NW Canadian continental margin (Max and Lowrie 1993).

East of the Mackenzie Delta (the Mackenzie River valley apparently did not provide a course for an ice stream) the margin is characterized by a number of major ice stream erosional troughs with north broadening mouths across northern Canada. Ice streams and fluvial systems have deeply incised the Amundsen Gulf and M'Clure Strait. These were major sediment flow courses from glaciated Canada into the Arctic Ocean. Further to the east, the continental margin to the north of the Queen Elizabeth Islands consists of many small north-flowing fjords and fluvial systems, and more restricted sediment fans. All of these troughs and the Mackenzie River Delta have pronounced continental margin sediment fans that have largely coalesced along this margin.

2.2 Greenland

This margin, where the base of the continental margin sediments rests on younger oceanic crust of the Eurasia Basin, was also heavily glaciated but has less continental margin sediment than to the west. Much less is known about it and the structural situation is unclear. The sedimentary wedge may be compressed because of impingement of Greenland onto the Atlantic margin of the Eurasia Basin as a result of opening of the Labrador Sea. If so, then deep and possibly thermogenic gas may be present in NGH along this margin. Max and Lowrie (1993) did not take this possible gas generation scenario into consideration; Eurasia Basin—Greenland continental margin sediments may have compound NGH.

2.3 Barents—Kara Seas Margin

This whole of this margin, along with its continental platform, was also heavily glaciated. But the history of continental margin sedimentation is not as extensive because the continental margins abut younger oceanic crust of the North Atlantic and Eurasia Basins. This zone extends from the SW margin of the Barents Sea to the 'corner' of the Eurasia Basin at which the relatively straight continental margin turns to the NW immediately to the SE of the islands of Severnaya Zemlya and the Taymyrskiy Peninsula. The boundary marks the passage from continental margin sediments with a glaciogenic origin from those to the east that are mainly fluvial in origin.

The western margin, between Norway and Svalbard is relatively straight and well defined, westerly-deepening shelf. The northern margin is relatively straight from a nearly right angle corner to the NW of Svalbard at the NW corner of the Barents Sea continental shelf to the Laptev Sea, but the shelf margin is locally irregular, with the deep St. Anna Trough separating Franz Joseph Land (Zemlya Frantsa Iosifa) from the easternmost shelf segment carrying the islands of Severnaya Zemlya and the mainland Taymyrskiy Peninsula.

During the last glacial maxima the NW European continental shelves, including the Barents and much of the northern Kara shelves, were ice covered (Polyak et al. 2008), but the extent of the ice varied considerably over the period from 100 to 90 kya to the end of the most recent glacial maxima at about 20 kya (Larsen et al. 2006). Ice cover prior to 100 kya may only be understood after the sedimentary sequences on the continental slopes are better known. Kleiber et al. (2001) show that during the last glacial maximum, large quantities of sediments were deposited directly upon the upper continental slope as the Kara Sea ice sheet advanced to the shelf break across almost the entire of the Barents-Kara sea shelves at 20 kya (Larsen et al. 2006). Under-ice detritus from accelerated weathering would have been deposited almost directly along the continental margins.

2.3.1 West Barents Margin

The Barents shelf has undergone extensive erosion, with the eroded sediments being deposited along the immediately adjacent continental margins. Vorren et al. (1991), whose quantification of geological processes on the Barents shelf has not been significantly revised, commented that throughout the Cenozoic, erosion of the shelf was tied to progradation of the shelf margin that from mid-Miocene to late Pliocene advanced the shelf margin between 20 and 40 km to the west. Uplift of the Barents shelf from mid-Pliocene to early Pleistocene resulted in a fluvial-glaciofluvial drainage system with very high sedimentation rates in the continental margin depocenters. Sediments in the Svalbard margin show various types of mass flows and turbidite sediments (Andersen et al. 1992), reflecting high sediment supply. Ingolfsson (2011) identifies five major trough-mouth sedimentary fans on this margin. Glacial ice stream erosion during the last 0.8 Ma was greatest in the SE Barents shelf than to the west and north. Gas venting near the crest of the Vestnesa Ridge on the west Barents continental margin has been tied to NGH in the sediments there (Vogt et al. 1994).

2.3.2 NW Barents Margin

This margin is marked by a well-defined continental margin between Svalbard and the islands of Franz Joseph Land, with other islands also set about the same distance back from the slope break to the continental margin. It carried a large ice cap during most of the most recent glacial episode. Hogan et al. (2010) and Ingolfsson (2011) note that ice stream directions were to the north, SE of a line passing to the SE from Svalbard. Cherkis et al. (1999) noted mass wasting in the continental margin sediments north of Svalbard that appeared to be tied to the rapid periodic formation of NGH and the development of subjacent low shear strength sediments that developed into near-slope parallel slip surfaces.

2.3.3 St. Anna Trough

The St. Anna Trough, which is the largest bathymetric depression in the Arctic continental shelves, bottoms below 500 and 1,000 m over a large area, extends north from Nova Zemlya to the shelf margin and provides a convenient geographic break between the Barents and Kara Seas. There appears to be a well-developed sediment fan across its entire mouth, which is probably the result of high sediment transport that has been carried into and along the trough.

2.3.4 Kara Sea—Eastern Margin

The Kara Sea and its shallow approaches to land to the east of Novaya Zemlya is a little over half the size of the Barents Sea but has a much shorter continental margin. Much of the northern Kara Sea was ice covered during those times that the Barents Sea ice cap extended to the east, at least during part of the last glaciation, which would have deflected river runoff from the Russian mainland, particularly the Pechora River. At times during at least the last glaciation, runoff that normally would have gone onto the Barents shelf was deflected by the Barents Sea icecap through the gap between the islands of Severnaya Zemlya and the mainland Taymyrskiy Peninsula (Fig. 1.1), the northernmost part of the Russian mainland. The incised channel is now largely sediment filled, mainly by clastic sediments (Gataullin et al. 2001), but its northern prolongation is a deep scour channel that broadens as it approaches the edge of the continental shelf.

Because this sediment would largely have been deposited along the geomorphologic North Barents margin, along with sediment that would have come through a gap in the islands of Severnaya Zemlya, the eastern margin of the zone is taken at the obvious knick point. Dittmers et al. (2004) shows from incised river channels that much of the Kara Sea was not ice covered for substantial periods. The Kara Sea also has proportionally more large rivers flowing into it from the Russian mainland, particularly the Ob and Yenisey Rivers, which would have brought a large sediment load, particularly during periods of major glacial ice melt. Much of this sediment load would have been transported north to the continental margin and the St. Anna trough. Other rivers, progressively closer to the shelf margin, flow off the Taymyrskiy Peninsula.

2.4 Laptev Sea—West Siberian Arctic Sea Zone

This is the shortest zone and extends across the almost orthogonal SE termination of the Eurasia Basin. Its eastern margin is the continental slope of the Siberian Amerasia Basin margin on which sediments were deposited prior to the opening of this continental margin segment. The crustal arch extending through the Novosibirskiy Islands to Henrietta and Bennett Islands nearer the margin of the East Siberian Sea shelf, probably separates two different groups of sediment delivery systems that may have resulted in two different depocenters along the continental margin of the East Siberian Sea, and this is taken as the margin with the East Siberian Sea zone.

The Siberian Shelf was not ice covered during the last glaciation, but the channel network from the existing mainland rivers is not well defined by erosional channels. In the northern Bering Sea and the continuation of the broad Siberian shelf to Alaska, an axial channel and a channel along the northwestern coast of Alaska may be mainly tidal scour channels in origin and not related to river runoff. Nonetheless, sediment erosion and winnowing could have taken place in them and they could also be associated with sands deposited on the continental slope.

This continental margin segment has a number of major rivers flowing onto the present shelf. The Lena River carries the largest sediment load and partly occupies the course of a more ancient river system that deposited at least Upper Cretaceous fluvial sediments in parts of its course. However, the present course of the lowermost Lena developed during the Pleistocene (Alekseev and Drouchtis 2004). In addition, the Khatanga, Yenisey, Olenka, and Yana Rivers flow onto the shelf.

2.5 East Siberian Sea Zone

This zone is the longest and along with the Laptev Sea contains one of the largest areas of very shallow continental shelf. A number of major rivers, particularly the Indigrirka, Kolyma, and Anadyr transport sediment from nearly Mongolia northward to the Arctic Ocean. Although possibly the least well known from the viewpoint of the history of river courses across it, the breadth and almost uniform slope of the shelf indicates that incised river systems have been filled and that the passage to sea level rise that covered the shelf probably involved large meandering river systems. The margin has existed more or less in its present location since the rifting and establishment of oceanic crust in the Amerasia Basin, although rotation of the Chukchi microcontinent (Grantz et al. 2011) may be associated with local uplift that could have affected river courses. We include a wide shelf area of the Chukchi Sea and the sea area off Alaska because we have no reason to distinguish it geologically or geomorphologically and the continental margin sediment thickness appears to be less abundant than in the North American zone to the east.

The boundary between zones 6.1 and 6.5, which is essentially that between glaciated and unglaciated hinterlands is placed to the west of the Mackenzie Delta and its submarine prolongation into the Canada Basin. Even though the Mackenzie may have been draining essentially unglaciated hinterland to the west, it was also draining the western margin of the North American ice cap to the east, which would have advanced and retreated with climate swings. The sediment from this margin was possibly reworked more extensively on land on marginal non-glaciated terrane before being deposited in the largest thick wedge of marine sediment in the Arctic Ocean, but in our view this only enhances the possibility of the presence of high grade NGH host sediments. Therefore, the boundary location more reflects the character of the sediment rather than the character of the immediate hinterland.

Apart from what appears to be the northward continuation on the shelf of the Indigirka River, shown by a reentrant of the 25 m contour (Jakobsson et al. 2004), there are few features that could allow picking of runoff and river channels from the present shoreline to the edge of the continental shelf.

The 2,600 + km long Kolyma River is the longest river draining eastern Siberia into the Arctic Ocean. It crosses part of the high Arctic that has not experienced Pleistocene lowland glaciation. The upper Kolyma on land in Siberia began to develop in the Cretaceous, when the main drainage divide between the Sea of Okhotsk and the Arctic Ocean formed (Patyk-Kara and Postolenko 2004). The modern Kolyma River also, in part, occupies the trend of much older river systems that drained this part of Siberia into the Arctic Ocean. During the Late Cretaceous and Cenozoic, two different processes affected the geology of the Siberian region: (1) compressional tectonics resulted in uplift associated with terrane accretion along the Northern Pacific subduction zone, and (2) opening of the Arctic Ocean Basin.

The Cenozoic history of the Lower Kolyma in Siberia and the continental shelf to the north in large part depends on the geological and geomorphological response to the opening of the Arctic Ocean Basin. Grantz et al. (2011) provide the most recent summary of the tectonic history of the Arctic Ocean region. The Siberian Shelf has been subjected to numerous transgressions and regressions that exposed much of the continental shelf to subaerial conditions (Patyk-Kara et al. 1980), although only the most recent sea level cycles related to sedimentation within present GHSZs are probably of importance for NGH exploration. The submerged course of the main drainage on the continental shelf has been identified for over 400 km out onto the continental shelf (Fig. 6 of Patyk-Kara and Postolenko 2004). Following the courses of these drainage channels that are now filled with dominantly coarser grained sediments may be useful to exploration near the shelf edge where water depths are deep enough to support GHSZ in the troughs and along the continental margins, where these river systems would be directly related to filled channels or to turbidite systems on the continental slope.

References

Alekseev MN, Drouchits VA (2004) Quaternary fluvial sediments in the Russian Arctic and Subarctic: late Cenozoic development of the Lena River system, Northeastern Siberia. Proc Geol Assoc 115:339–346

Andersen ES, Solheim A, Elverhoi A (1992) Development of a glaciated Arctic continental margin: exemplified by the western margin of Svalbard. Proceedings international conference on Arctic margins OCS study, pp 155–160

Cherkis NZ, Max MD, Vogt PR, Crane K, Midthassel A, Sundvor E (1999) Large-scale mass wasting on the North Spitsbergen continental margin, Arctic Ocean. In: Gardiner J, Vogt P, Crane K (eds) Mass wasting in the Arctic, vol 19. Geomarine Letters Special Issue, pp 131–142

Dittmers K, Niesen F, Stein R (2004) Fluvial history of an ice sheet proximal continental shelf: The Southern Kara Sea, West Siberia during late quaternary. (Abs) European Geosciences Union First General Assembly, Nice, Apr 2004, p 1, hdl: 10013/epic.2204

Egawa K, Furukawa T, Saeki T, Suzuki K, Narita H (2013) Three-dimensional paleomorphologic reconstruction and turbidite distribution prediction revealing a Pleistocene confined basin system in the northeast Nankai Trough area. AAPG Bull 97(5):781–798. doi:10.1306/10161212014

Frye M, Shedd W, Boswell R (2011) Gas hydrate resource potential in the Terrebonne Basin, Northern Gulf of Mexico. Mar Pet Geol 34:19. doi:10.1016/j.marpetgeo.2011.08.001

Gataullin V, Mangerud J, Svendsen JI (2001) The extend of the late Weichselian ice sheet in the Southeastern Barents Sea. Global Planet Change 31:453–474

Grantz A, Hart PE, Childers VA (2011) Development of the Amerasia and Canadian Basins, Arctic Ocean. In: Spenser AM, Embry AF, Gautier DL, Stompkova AV, Sorensen K (eds) Arctic petroleum geology, vol 35. Geological Society of London Memoir, pp 771–799. doi:10.1144/M35.50

Hogan KA, Dowdeswell JA, Noormets R, Evans J, O'Cofaigh C, Jakobsson M (2010) Submarine landforms and ice-sheet flow in the Kvitoya Trough, Northwestern Barents Sea. Quat Sci Rev 29(25–26):3545–3562. doi:10.1016/j.quasirev.2010.08.015

Ingolfsson O (2011) Fingerprints of quaternary glaciations on Svalbard. Geol Soc London Spec Publ 354:15–31. doi:10.1144/sP354.2

Jakobsson M, Mcnab R, Cherkis N, Shenke H-W (2004) The international map of the Arctic ocean (IBCAO). Polar stereographic projection, scale 1:6,000,000. Research publication RP-2. U.S. National Physical Data Center, Boulder, Colorado 90305

Jakobsson M, Macnab R, Mayer L, Anderson R, Edwards M, Hatzky J, Schenke H-W, Johnson P (2008) An improved bathymetric portrayal of the Arctic Ocean: implications for ocean modeling and geological, geophysical and oceanographic analyses. Geophys Res Lett 35(5):L07602. doi:10.1029/2008GL033520

Kleiber HP, Niessen F, Weiel D (2001) The late quaternary evolution of the Western Laptev Sea continental margin, Arctic Siberia—implications from sub-bottom profiling. Global Planet Change 31:105–124

Larsen E, Jjaer KH, Demidov N, Funder S, Grosfjeld K, Houmark-Nielsen M, Jensen M, Linge H, Lysa A (2006) Late pleistocene glacial and lake history of Northwestern Russia. Boreas 35:31. ISSN 0300-9483. doi:10.1080/03009480600781958

Max MD, Lowrie A (1993) Natural gas hydrates: Arctic and Nordic Sea potential. In: Vorren TO, Bergsager E, Dahl-Stamnes ØA, Holter E, Johansen B, Lie E, Lund TB (eds) Arctic geology and petroleum potential. Proceedings of the Norwegian petroleum society conference, 15–17 Aug 1990, Tromsø, Norway. Norwegian Petroleum Society (NPF), special publication 2 Elsevier, Amsterdam, pp 27–53

Patyk-Kara NG, Postolenko GA (2004) Structure and Cenozoic evolution of the Kolyma river valley: from upper reaches to continental shelf. Proc Geol Assoc 115:325–338

Patyk-Kara NG, Morozova LN, Biryukov VY, Novikov VN (1980) New data on the structural-geomorphological setting of coastal plains and shelf of East Arctic Seas. Geomorfologiya 3:9–98

Polyak L, Niessen F, Gataullin V, Gainanov V (2008) The eastern extent of the Barents-Kara ice sheet during the last glacial maximum based on seismic-reflection data from the eastern Kara Sea. Polar Res 27:162–174. doi:10.1111/j.1751-8369.2008.00061.x

Vogt PR, Crane K, Sundvor E, Max MD, Pfirman SL (1994) Methane-generated (?) pockmarks on young, thickly sedimented oceanic crust in the Arctic: Vestnesa Ridge, Fram Strait. Geology 22:255–258

Vorren T, Richardsen G, Knutsen S, Henriksen E (1991) Cenozoic erosion and sedimentation in the Western Barents Sea. Mar Pet Geol 8:317–340. doi:10.1016/0264-8172(91)90086-G

Chapter 3
Natural Gas Hydrate: Environmentally Responsive Sequestration of Natural Gas

Abstract Pleistocene glacial sediments will predominantly host NGH, and possibly only those to depths of no more than 1 km. Older sediments will likely be buried too deeply to host NGH. Each glacial episode would have produced a suite of sediments related to the onset and glacial maximum period and especially during the onset of the following interglacial period when the melting of the ice cap would produce large volumes of water that would strongly affect sediment winnowing and transport. These periods of maximum water flow would be likely to produce the clastic sandy sediments that would be ideal hosts of high-grade NGH deposits in the deeper continental shelves and the continental slopes. In addition, sea level variation would have strongly controlled the position of the shoreline positions during the glacial and interglacial cycles. Sediments within about 1.2 km depth below the seafloor comprise the exploration zone for NGH and related gas deposits. The continental margins of the Arctic Ocean have been divided into 5 regions for analysis of the degree to which they could provide optimal NGH host sediments to suitable depositional environments.

Keywords Barents • Laptev • East siberian • Alaska • Canada • Greenland • Sediment host • Natural gas hydrate • NGH

NGH is a crystalline material composed of water molecules that form cage structures and gas molecules that occupy almost all of the cages, the whole being stabilized by the weak electrical force, van der Waals bonding. NGH is directly analogous to metallic and other economic mineral deposits formed by crystallization from mineralizing solutions (Max et al. 2006; Sloan and Koh 2008; Demirbas 2010; Giavanini and Hester 2011). This differentiates it strongly from gaseous and liquid (including tar sands) hydrocarbon deposits. Naturally occurring NGHs are composed predominantly of methane, and NGH is often referred to as 'methane hydrate even in the absence of analyses, but appreciable heavier hydrocarbon gases such as ethane, propane, and butane may occur where thermogenic gases have contributed to the gas flux. Because the crystalline cage structures closely pack the gas molecules, one m^3 of Structure I methane NGH will produce about 160 m^3 of methane (at STP), equivalent to an energy density of over 165,000 BTU/ft^3. This compression factor, regardless of seafloor depth and ambient pressure, is a principal component of the economic value system for NGH. Compound NGH, formed from mixed or higher density natural gases, will have less gas per m^3 than methane-NGH but increased BTU content determined by the gas mixture. Compound NGH formed from a mixture of gases is more stable (at lower pressures and higher temperatures) than pure methane-NGH.

NGH forms spontaneously when the right combinations of elevated pressure and low temperature conditions exist, and when a suitable concentration of hydrate-forming gas can mix with and react with water (Fig. 3.1a). NGH is commonly found within a gas hydrate stability zone (GHSZ) that extends from near a cold surface on land or the seafloor downward to some depth at which increasing temperature renders NGH unstable (Fig. 3.1b). Although NGH is generally stable in the ocean water at great depths (below about 700 m in the example of Fig. 3.1b), it will not survive there because it will float upward due to its low density (about 0.9) and will dissociate when it passes through the phase boundary. The temperature of the water column in the open ocean decreases with depth as a result of the cold, dense, saline water produced as a byproduct of freezing ice.

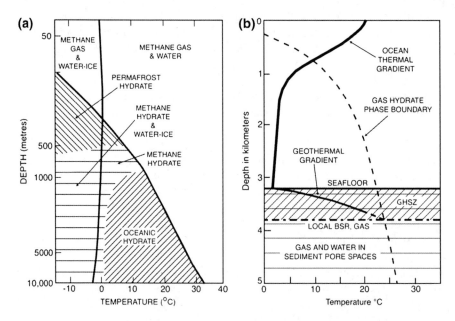

Fig. 3.1 a (*left*) Generalized phase boundary for methane hydrate and economic NGH zones. *Solid line, water ice.* **b** (*right*) Oceanic GHSZ as a function of typical open ocean thermal gradient and geothermal gradients. Original figures after Kvenvolden (1988) and Max (2003)

In the Arctic Ocean, the water temperature at depth is at or below 0 °C, which places the minimum water depth for NGH stability at about 250 m water depth. We prefer to assume a depth of about 300–350 m for the minimum depth of a continuous or year-round GHSZ in the Arctic Ocean Basin sediments. As the water depth increases, and pressure increases, the GHSZ increases in thickness. In general, thicker GHSZ sections are more likely to have more significant NGH concentrations. In addition, geothermal gradients, which describe the average change in temperature with depth over specific intervals, may vary considerably.

Oceanic NGH differs from all other types of natural gas deposits in being responsive to changes in its environment. The greater part of the hydrocarbon gases in NGH appears to have been generated at depth from organic matter either biogenically or thermogenically and has been transported upward into the GHSZ, driven by tectonic forces or dewatering of the sediment pile driven by compaction. High TOC sediments can also generate considerable local gas, but TOC-rich sediments tend to be muddy and have poorly defined permeability paths. In these, NGH-forming gas moves primarily by diffusion in the groundmass or by flow in fractures. Deep sources are particularly obvious when the gas has a thermogenic signature containing a mixture of higher-density hydrocarbon gases.

NGH may occur at any level within the GHSZ, depending on the nature of the 'plumbing system' utilized by groundwater and natural gas to bring the mineralizing solutions into the GHSZ from the sediment pile. High-grade, but more broadly distributed and deeper NGH concentrations, in which the mineralization is directly related to gas transported by water movement into permeable strata in the GHSZ suggests that mineralizing solutions migrated along porous geological horizons that passed up into the GHSZ. Higher level and isolated concentrations suggest that faults or fracture systems, along which water flow may be much faster, are able to bring the mineralizing solutions higher in the GHSZ before crystallizing in a favorable host strata or vein work. For instance, fractured shale carrying substantial NGH in vein works occurs in close proximity to sand-hosted NGH concentrations in the northern GoM (Boswell et al. 2011).

When the gas reaches the GHSZ, in which pressure and temperature are suitable for NGH formation (Fig. 1.1), NGH will form so long as the gas concentration in the pore water media is high enough. As the dissolved gas concentration rises and falls, the hydrate growth dynamic alters and greater or lesser amounts of hydrate may form. Free gas is usually associated with groundwater that is saturated or supersaturated with dissolved gas. If the dissolved gas concentration falls below a certain level, usually identified as the vapor pressure of the hydrate-forming gas in the hydrate, the hydrate will dissolve. Thus, NGH is not a material that is fixed to a particular geological attribute over time. It occurs only within a GHSZ, and as that may change thickness or position depending on sedimentary deposition or erosion, the presence of NGH within it also will migrate from an area in which NGH is unstable and into an area where it is stable, as may happen when sea level rises and pressure increases, or sediment erosion takes place. NGH may reposition itself by a process of recrystallization.

NGH is very environmentally responsive, being able to crystallize, dissociate and dissolve, migrate, and recrystallize when environmental conditions change. Under conditions of continued sedimentation, for instance, the base of the GHSZ will move upwards because it is keyed to the sediment surface. The balance between the seafloor low temperature and the rising temperature with depth is maintained. NGH that was once within the GHSZ at its base will convert naturally to gas when it passes out of the field of NGH stability and either pond or tend to rise into the GHSZ where it can form more NGH. A fall in sea level will also lower pressure on the seafloor, causing the base of GHSZ to rise. In contrast, sediment erosion will cause the temperature gradient to move deeper into the sediment and lower the base of the GHSZ, as will rise in sea level that raises the pressure. When NGH migrates with the GHSZ in order to maintain its thermodynamic stability, it may leave behind virtually no definitive evidence of its one-time presence.

Seafloor warming will cause thinning of the GHSZ while seafloor cooling will cause thickening: these are the environmental changes usually associated with methane release or greater sequestration. Changes in pore water salinity or chemical composition, while probably relatively rare, have the potential to move the position of the NGH phase boundary. Concentration, of course, is key; if the concentration of the dissolved NGH-forming gas falls sufficiently, dissolution takes place.

Four general classes of NGH in different surroundings have been identified (Table 3.1). These classes are based on thermodynamic models that were used to estimate the costs of processes that are required to convert NGH to its constituent gas and water (Moridis and Kowalsky 2006). There are subtle but important differences in the petroleum systems related to each of these classes of NGH concentration. Note that the presence of a geological trap is not one of the criteria for any NGH class.

NGH concentrations that may prove economic may only be found in a GHSZ thick enough to host significant concentrations with enough overlying sediment to stabilize the reservoir by its weight during drilling, NGH conversion, and gas extraction. Depending on the sediments within the GHSZ and the presence and disposition of NGH, there is a wide range of geotechnical possibilities. For instance, the presence of multiple NGH-enriched units that can be mechanically very strong when NGH concentrations are high, essentially binding the sediment grains into a solid aggregate or where sediment grains may be cemented together

Table 3.1 NGH classification

Class	NGH	Bounded	Materials in contact	Geological situation	System
1	Concentrated	Permeability boundaries/ geological strata	Gas (over water)	1a Oceanic	Open
				1b Oceanic, Permafrost	Closed
2			Mobile water	Oceanic	Open
3			No gas or water	Dry gas trap (including vein-type NGH)	Closed
4	Dispersed	Few permeability boundaries	Pore water	Fine-grained marine sediments	Very open

Revised from Moridis and Kowalsky (2006)

by thin NGH films, can affect overall sediment stability. The minimum thickness of a GHSZ necessary to host a prospective producing NGH deposit is not known. An NGH prospect and its host and overlying sediments will have to undergo geotechnical analysis as part of developing a production plan. With time and experience, less conservative safety margins will probably emerge. Although it is not known to what water depths possible NGH concentrations may occur, there will probably be some maximum depth below which exploration cutoffs will apply for either operational considerations or some aspect of NGH paragenesis.

Ideally, large concentrations of NGH are most likely to be found near the base of relatively thick GHSZs in which the sediment will be more compacted under the influence of gravity than in shallower sediments with thinner GHSZ. The more compacted nature of the NGH host sediments and particularly the degree of compaction and reduced permeability of sediments bounding a sand-hosted NGH deposit may strongly control the percentage of technically and economically recoverable natural gas. NGH will generally occur world wide in semi-compacted sediments that do not have the geologically strong character of conventional natural gas reservoirs. The stronger the geomechanical character of the reservoir sediments and between the NGH deposit and the seafloor, the lower the risk of mechanical failure and more likely higher rates of NGH conversion and natural gas production.

References

Boswell R, Collett TS, Frye M, McConnell D, Shedd W, Dufrene R, Godfriaux P, Mrozewski S, Guerin G, Cook A (2011) Gulf of Mexico gas hydrate joint industry project leg II: technical summary. US Department of Energy, NETL, Morgantown

Demirbas A (2010) Methane gas hydrate. Springer Geophysics, p 287. ISBN 978-1-84882-872-8

Giavarini C, Hester K (2011) Gas hydrates: immense energy potential and environmental challenges, green energy and technology series. Springer, ISBN-13: 978-0857299550, p 160

Kvenvolden KA (1988) Methane hydrate—a major reservoir of carbon in the shallow geosphere? Chem Geol 71:41–51

Max MD, Johnson A, Dillon WP (2006) Economic geology of natural gas hydrate. Springer, Berlin, Dordrecht, p 341

Max MD (ed) (2003) Natural gas hydrate: in oceanic and permafrost environments, 2nd edn. Kluwer Academic Publishers (now Springer), London, Boston, Dordrecht, p 422

Moridis GJ, Kowalsky M (2006) Gas production from unconfined class 2 oceanic hydrate accumulations. In: Max MD (ed) 2003 Natural gas hydrate: in oceanic and permafrost environments, 2nd edn. Kluwer Academic Publishers (now Springer), London, Boston, Dordrecht, pp 249–266

Sloan ED, Koh CA (2008) Clathrate hydrates of natural gases, 3rd edn. Taylor & Francis/CRC Press, p 720

Chapter 4
NGH as an Unconventional Energy Resource

Abstract NGH is one of a number of unconventional gas plays. It is essentially stable within its reservoir, and as such has an extremely low environmental risk for blowout so long as prudent and very simple and inexpensive exploration and production methodologies are followed. Pressure, temperature or groundwater chemistry can be altered to achieve dissociation. In addition, conversion can be induced through a controlled dissolution process. Although gas resource-rich countries such as Russia, Canada, and the United States have suspended or reduced their research in NGH as an energy resource, NGH may be important in climate change modeling and remains the subject of study. In countries that have little indigenous energy resources, political concerns related to obtaining secure local gas resources might be more important to NGH development than the delivered gas price of imported resources. Near term development plans for NGH including drilling and gas production tests are being pursued, particularly in Japan, India, Korea, and possibly China.

Keywords Thermodynamic trap • Large amounts • Natural gas hydrate • GHSZ • Gas hydrate stability zone • Oceanic • Permafrost • NGH

A conventional reservoir is naturally pressurized, or more commonly, over-pressurized for the formation pressure of a hydrocarbon-free sediment at the same depth and geological situation. Conventional gas deposits will flow spontaneously to the surface when a reservoir is drilled, and the risk of blowout venting is pervasive in all conventional hydrocarbons, particularly those found at greater depths and higher pressures.

In contrast, an unconventional deposit is one that must be stimulated in some way in order to cause the hydrocarbon to flow. In a conventional gas deposit these processes (steam or hot water injection, chemical solvent injection, gas injection, etc.) would be considered to be a secondary recovery technique that are applied to recover hydrocarbons that would otherwise have to be left in the reservoir.

NGH is one of a number of unconventional gas plays that is essentially stable within its reservoir. The physical conditions of unconventional gas plays must be altered in some way to allow the natural gas to be produced. In the case of coal-bed methane, water is pumped out of the gas-infused coal-shale measures, resulting in the release of largely dissolved gas. In the case of shale gas and tight gas sands, permeability has to be induced by fracking. In the case of NGH, the physical conditions affecting its stability field (i.e., Fig. 1.1) have to be altered so that the NGH converts to its gas and water components. Either or both pressure or temperature can be altered to achieve the conversion. In addition, conversion can be induced through a controlled dissolution process (Max et al. 2006) and molecular substitution.

NGH exists in a very different way to other gas deposits. Conventional gas and the other unconventional gas deposits are related to geological traps in which they may have resided for hundreds of millions of years. The other unconventional gas deposits also have considerable geological permanence. In contrast, oceanic NGH resides in an existing thermodynamic or physical/chemical 'trap' that is strongly influenced by changing environmental conditions. Natural gas in solid NGH is the result of a crystallization process driven by the mineralizing solutions within the GHSZ. In fact, type 2 concentrations (Moridis and Kowalsky 2006) may occur in an open pore water situation in which subjacent water passes through the NGH-enriched zone and may vent to the seafloor. If the water media in contact with the NGH ceased having sufficient gas concentration, the NGH would dissolve into the water to maintain diffusional balance.

Perhaps the principle reason why there is wide interest in NGH is that there appears to be huge volumes of natural gas sequestered in NGH, and large concentrations have been identified. Estimates of natural gas in NGH indicate that the resource base may be of greater volume than is estimated to be in conventional gas deposits (Boswell and Collett 2011). In order to optimize exploration for oceanic NGH, however, it is important to understand the paragenesis and geological model for concentrated NGH, so that exploration can be focused upon them. NGH occurs in permafrost regions and in oceanic marine sediments in deep continental shelves and margins (Brown et al. 1997; Max et al. 2003).

4.1 Permafrost NGH

Permafrost regions (Fig. 4.1) may host considerable NGH as part of a compound ice-NGH cryosphere (Max et al. 2006), having water ice stable from the base of the active surface layer that melts in the summer and refreezes in the winter, downward to just above the 0 °C isotherm. NGH is stable from about 200–225 m below the surface for methane NGH and shallower for compound NGH (Oberman and Kakunov 1978). The base of the GHSZ is deeper than that of ice permafrost. Because of pressures at depth, the base of the GHSZ may be stable

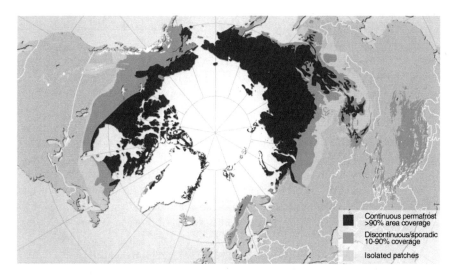

Fig. 4.1 Permafrost in the northern hemisphere. From NSIDC (2012). Original reference Brown et al. (1997)

to temperatures commonly from 6 to 10 °C in permafrost terrane, but because of varying geothermal gradients, but depth variation of GHSZ thickness can occur, often on a relatively small lateral scale., In addition, average surface temperature variation will also affect GHSZ thickness.

4.1.1 Geologically Trapped

Concentrations of NGH were first identified in the higher elevations of Siberian gas fields associated with thick permafrost terrain (Makogon et al. 1972). NGH was then identified in similar situations on the Alaskan North Slope and in the Mackenzie Delta (Collett et al. 1999). These discoveries were a byproduct of conventional hydrocarbon exploration. The NGH generally occurs in the upper part of conventional gas traps along with subjacent conventional gas residing below the GHSZ. The gas was apparently trapped before the NGH was formed (Collett 1995, 1997, 2002; Collett et al. 1999, 2008) and converted to NGH as the cold conditions of glacial episodes penetrated into the ground (Max et al. 2006). These deposits are essentially just a special type of conventional gas deposit. Some permafrost NGH deposits may also exist beneath the present continental shelves; they may be widespread in permafrost regions. Conventional petroleum analysis has been used to explore for these deposits and will continue to be. Natural conversion of the NGH in these deposits simply results in more free gas in the reservoir; little or none of which can vent to the atmosphere unless structural activity opens fracture systems.

4.1.2 Permafrost-Related, Non-Geologically Trapped

NGH in the compound cryosphere may occur in suitable host strata without being constrained by geological traps and without natural gas having been concentrated before the formation of NGH from it. From an exploration point of view, these deposits will share many of the attributes of oceanic NGH paragenesis and have a strongly related petroleum system, but it is unlikely that the rocks and sediments in a permafrost area GHSZ will have the same weak geotechnical character and permeability characteristics of shallow marine sediments, except where NGH forms in relatively young sediments.

The compound ice-water cryosphere extends out beneath shallow continental shelves, potentially down to present water depths of about 120 m that were exposed during glacial maxima (Collett and Dallimore 2003). Both water ice and NGH may occupy large shallow shelf areas that did not have ice caps. The sea level lowstand of about -120 m is about consistent with worldwide sea level lowstands (Lambeck 2004; Long et al. 2008). Because ice melts at 0 °C and permafrost melts generally from the surface downward, NGH may persist in continental shelves after all the integral and superjacent ice has melted because it is stable at higher temperatures. The extent of NGH both in onshore and subsea permafrost NGH is very poorly known but there are huge amounts of venting methane from the East Siberian Sea, much of it presumed to be derived from converting NGH (Shakhova and Semiletov 2007, 2010). Subsea permafrost itself has locally been identified in the southern Kara Sea (Rekant et al. 2005). Much of the gas produced by natural conversion of NGH from shelf areas can be expected to vent to the seafloor where it will likely dissolve in the seawater with relatively little reaching the atmosphere.

4.1.3 Vein-Type Deposits

There is a possibility that vein-type NGH similar to the recent discovery of this paratype in the shallow permafrost terrain of the Qinghai-Tibet Plateau in western China (Lu et al. 2010) may be much more widely developed in permafrost terrane. Drilling appears to be the primary exploration method for vein type NGH deposits in ice rich terrane. NGH and water ice have similar responses to acoustic and electrical exploration methods. The petroleum system is described by Max and Johnson (2011) with respect to a direct formation by a chemical reaction between water ice and natural gas, but we do not regard vein-type deposits to be of immediate interest for economic development. However, these deposits may be of importance to climate change. Where vein-type NGH is dissociated by warming of the Arctic region, which is currently causing permafrost to destabilize, gas from the NGH may find its way directly into the atmosphere with much more facility than any other type of NGH deposit.

Because permafrost NGH concentrations are either similar in their provenance and petrogenesis to conventional gas deposits, of which they constitute a special type, and other permafrost NGHs are very poorly known, permafrost NGH plays, including the relic subsea permafrost on continental shelves, are not discussed here further.

4.2 Oceanic NGH

We focus in this publication mainly on oceanic NGH, which may have significant economic potential in single deposits or fields. In fact, the Japanese Nankai deposit has already been raised to prospect status (JOGMEC 2013). In particular, however, we focus on the likelihood and location of oceanic High Arctic NGH potential.

It is probably that commercial quantities of oceanic NGH may occur either dispersed or in fracture zones in muddy sediment reservoirs (McGee et al. 2009) in which drilling results indicate that substantial quantities of NGH occur Frye 2011). However, they are not regarded as primary exploration targets, as exploration for them and production from them is uncertain because of reservoir instability during NGH conversion and a production concept or model is lacking (Boswell and Collett 2011). Although most NGH occurs dispersed or in veinworks in muddy sediments, the greatest concentrations of NGH that have economic potential occur in sands and coarse-grained sedimentary strata (Max et al. 2006; Boswell and Collett 2011). Thus, exploration is first for sand hosts that are similar in many ways to conventional gas deposits (Max et al. 2006). Industry is already skilled in this facet of exploration. Without suitable host sediments and pore water activity, large volumes of concentrated NGH are unlikely to occur.

Oceanic NGH is the only NGH option for countries with little or no permafrost terrane. Nations with major expanses of permafrost, the United States, Canada, and the Russian Federation are all energy-rich. But virtually the rest of the world has an interest in identifying and exploring their NGH potential. Exploration for oceanic NGH has made remarkable strides in the last 20 years, particularly in the national programs of Japan (Noguchi et al. 2011), the United States (Boswell et al. 2011), Canada (Dallimore and Collett 2005), and India and Korea (Long et al. 2008). Although the NGH programs of Canada and the United States were primarily driven by economic interest 10 years ago, newly proven resources, including shale gas and oil and tar and heavy oil sands, have reduced the imperative to develop NGH around North America as a near-term gas resource. In countries that have few indigenous hydrocarbon energy resources, however, political concerns related to obtaining secure local gas resources may be more important to NGH development than the availability of imported gas.

References

Boswell R, Collett TS (2011) Current perspectives on gas hydrate resources. Energy Environ Sci 4:1206–1215. doi:10.1039/c0ee00203h

Boswell R, Collett TS, Frye M, McConnell D, Shedd W, Dufrene R, Godfriaux P, Mrozewski S, Guerin G, Cook A (2011) Gulf of Mexico gas hydrate joint industry project leg II: technical summary. US Department of Energy, NETL, Morgantown

Brown J, Ferrians OJ, Heginbottom JA, Melnikov ES (1997) Circum-Arctic map of permafrost and ground ice conditions. International Permafrost Association. U.S. Geological Survey Circum-Pacific Map Series Map CP-45

Collett TS, Agena WF, Lee MW, Zyrianova MVKJ, Charpentier RR, Troy Cook T, Houseknecht DW, Klett RR, Richard M, Pollastro RMJ, Schenk CJ (2008) Gas hydrate resource assessment, North Slope, Alaska. USGS fact sheet from October 2008, p 3. http://geology.com/usgs/alaska-gas-hydrates.shtml

Collett TS, Dallimore SR (2003) Permafrost-associated gas hydrate. In: Max MD (ed) Natural gas hydrate in oceanic and permafrost environments. (Kluwer) Springer, Berlin, pp 43–60

Collett TS (2002) Energy resource potential of natural gas hydrates. Am Assoc Pet Geol Bull 86(11):1971–1992

Collett TS (1997) Gas hydrate resources of Northern Alaska. Bull Can Petrol Geol 45:317–338

Collett TS (1995) Gas hydrate resources of the United States. In: Gautier DL, Dolton GL, Takahashi KI, Varnes KL (eds) National assessment of United States oil and gas resources-results, methodology, and supporting data. U.S. geological survey digital data series 30 (on CD-ROM)

Collett TS, Lewis RE, Dallimore SR, Lee MW, Mroz TH, Uchida T (1999) Detailed evaluation of gas hydrate reservoir properties using JAPEX/JNOC/GSC Mallik 2L-38 gas hydrate research well downhole well-log displays. In: Dallimore SR, Uchida T, Collett TS (eds) Scientific results from JAPEX/JNOC/GSC Mallik 2L-38 Gas hydrate research well, Mackenzie Delta, Northwest Territories, Canada. Geol Surv Can Bull 544:295–312

Dallimore S, Collett T (2005) Scientific results from the Mallik 2002 gas hydrate production research well program, Mackenzie delta, Northwest Territories, Canada. GSC Bulletin 585

Frye M, Shedd W, Boswell R (2011) Gas hydrate resource potential in the Terrebonne Basin, Northern Gulf of Mexico. Mar Pet Geol 34:19. doi:10.1016/j.marpetgeo.2011.08.001

JOGMEC (2013) News release. Gas production from Methane hydrate layers confirmed, p 3. www.jogmec.go.jp

Lambeck CR (2004) Sea-level change through the last glacial cycle: geophysical, glaciological and palaeogeographic consequences. CR Geosci 336:677–689

Long PE, Wurstner SK, Sullivan EC, Schaef HT, Bradley DJ (2008) Preliminary geospatial analysis of arctic ocean hydrocarbon resources. U.S. Department of Energy/Pacific Northwest National Laboratory PNNL-17922

Lu Z, Zhu Y, Zhang Y, Wen H, Li Y, Jia Z, Liu C, Wang P, Li Q (2010) Gas Hydrate features in the Qilian Mountain permafrost, Qinghai Province, China. In: Boswell R (ed) Fire in the Ice, March 2010. National Energy Technology Center, Morgantown, pp 6–9

Makogon YF, Trebin FA, Trofimuk AA, Tsarev VP, Cherskiy NV (1972) Detection of a pool of natural gas in a solid (hydrated gas) state (in English): Doklady Akademii Nauk SSSR, vol 196. pp 203–206; Doklady-Earth science section, vol 196. pp 197–200

Max MD, Johnson AH (2011) Diagenetic methane hydrate formation in permafrost: a new gas play? 2011. Offshore technology conference. In: Proceedings, OTC arctic technology conference, Houston, Texas, USA, 7–9 Feb 2011, p 7

Max MD, Johnson A, Dillon WP (2006) Economic geology of natural gas Hydrate. Springer, Berlin, Dordrecht, p 341

Max MD, Jürgen Mienert J, Andreassen K, Berndt C (2003) Gas hydrate in the Arctic and Northern North Atlantic Oceans. In: Max MD (ed) 2003 Natural gas hydrate: in oceanic and permafrost environments, 2nd edn. Kluwer Academic Publishers (now Springer), London, Boston, Dordrecht, pp 171–182

References

McGee T, Lutken C, Woolsey JR, Rogers R, Dearman J, Brunner JC, Lynch FC (2009) Can fractures in soft sediments host significant quantities of gas hydrates? In: Collett TS, Johnson AH, Knapp C, Boswell R (eds) Natural gas hydrates—energy resource potential and associated geologic hazards. Am Assoc Petrol Geol Mem 89:297–307

Moridis GJ and Kowalsky M (2006) Gas production from unconfined class 2 oceanic hydrate accumulations. In: Max MD (ed) 2003 Natural gas hydrate: in oceanic and permafrost environments, 2nd edn. Kluwer Academic Publishers (now Springer), London, Boston, Dordrecht, pp 249–266

Noguchi S, Furukawa T, Aung TT, Oikawa N (2011) Reservoir architecture of methane hydrate bearing turbidite channels in the eastern Nankai Trough, Japan. In: Proceedings of the 7th international conference on gas hydrates (ICGH 2011), Edinburgh, Scotland, United Kingdom, 17–21 July 2011, p 9

NSIDC (2012) State of the cryosphere. National snow & ice data center. www.nsidc.org/cryosphere/sotc/permafrost.html

Oberman NG, Kakunov BB (1978) Determination of the thickness of permafrost on the Arctic coast. Permafrost: second international conference USSR contribution. U.S. National Academy of Sciences, pp 143–176

Rekant P, Cherkashev G, Vanstein B, Krinitsky P (2005) Submarine permafrost in the nearshore zone of the southwestern Kara Sea. Geo-Mar Lett 25:183–189

Shakhova N, Semiletov I (2010) Extended abstract and presentation. Methane release from the East Siberian Arctic Shelf and the potential for abrupt climate change. SERDP / ESTCP partners in environmental technology. Technical session 1A. Technical symposium and workshop. Marriott Wardman Park Hotel, Washington, DC, Nov 30–2 Dec 2010

Shakhova N, Semiletov I (2007) Methane release and coastal environment in the East Siberian Arctic shelf. J Mar Syst 66:227–243. doi:10.1016/j.jmarsys.2006.06.006

Chapter 5
Elements of the NGH Petroleum System

Abstract NGH is essentially a modern hydrocarbon mineral deposit in which existing NGH concentrations are in thermodynamic equilibrium with their surroundings. The objective of NGH petroleum system analysis is the same as it is for conventional petroleum system analysis, that is, to provide a methodology for hydrocarbon concentrations and to identify potentially commercially producible sources of natural gas. NGH concentrations reflect the convergence of a number of existing conditions rather than a succession of geological conditions, the timing of which were critical to the formation of conventional gas and petroleum deposits. The elements of the NGH petroleum system consist of: (1) Sufficient gas flux, (2) Migration pathways from subjacent sources toward the seafloor, (3) High-grade host reservoir sediments, (5) Suitably thick GHSZ. A number of parameters of conventional petroleum system analysis are not necessary for analysis of the NGH system. These include the long-term history of the basin, thermal history of sediments, multiple petroleum systems in a basin, and detailed stratigraphic analysis much below the present GHSZ.

Keywords Unconventional • Modern • Natural gas hydrate • GHSZ • Gas hydrate stability zone • Transient • System • NGH

NGH petroleum system analysis is much more direct than conventional petroleum system analysis, and is almost certainly a good deal less expensive to carry out, as much less information is required. NGH petroleum system analysis follows the general model of conventional hydrocarbon systems but has significant differences. The objective of NGH petroleum system analysis is the same as it is for conventional petroleum system analysis, that is, to provide a methodology for hydrocarbon concentrations and to identify commercial sources of natural gas (Max and Johnson 2013). In the NGH petroleum system there is no need to link reservoirs with specific source beds, there is no need to understand the entire thermal history of the basin, and most important, there is no need to identify a geological trap. NGH concentrations reflect the convergence of a number of existing conditions rather than a succession of geological conditions, the timing of which were critical to the formation of conventional gas and petroleum deposits.

NGH is essentially a 'modern' deposit in which existing NGH concentrations are in thermodynamic equilibrium with their surroundings (Max and Johnson 2011). Should the surrounding conditions of temperature, pressure Natural Gas Hydrate (NGH): Pressure, or gas concentration in surrounding media change, the NGH will respond to these changes, slowed only by the natural buffering inherent in NGH formation and dissociation. NGH concentrations can change their geological position, whereas conventional hydrocarbon deposits remain trapped in their reservoirs unless breached by faults or other leakage. NGH concentrations are tied to subsurface conditions and may not have persisted long in their existing geological setting.

NGH petroleum system analysis focuses on the GHSZ as NGH only occurs within it. The GHSZ and the immediately subjacent porous horizons (that may contain free gas) and groundwater (pore water) feeder systems are the only part of the entire sediment succession that is important to the NGH petroleum system. Further information about deeper sediments and structure may enhance feeder system information but will not comment directly upon NGH concentrations. Exploration may be confined to the upper 1–2 km of marine sediments, and probably only in water depths greater than about 600 m. The porous strata for some distance below the GHSZ may have free gas.

An excellent example of conventional petroleum system analysis carried out in the western end of the northern GoM (Fiduk et al. 1999) demonstrates the time-sensitive relationships between petroleum generation, migration, and concentration in geological traps. In contrast, NGH is essentially a more 'modern' deposit. The NGH petroleum system is relatively uncomplicated (Fig. 5.1). It consists of relatively few basic elements, all of which have to be active or interactive now and in the recent geological past (Max and Johnson 2013). Natural gas has to be produced either biogenically or thermogenically at locations from where it could rise toward the surface and reach the GHSZ. Migration pathways, either along permeable beds or faults, or a combination of them are needed to connect the gas sources with the GHSZ. There has to be a GHSZ thick enough to sustain a continuity of

Fig. 5.1 Schematic diagram of the NGH petroleum system. The table format is similar to that of conventional hydrocarbon system tables, but much simpler

Sediment Host of Any Geological Age	NGH Petroleum System
Yes/No	Hydrate Concentrations
Yes/No	Reservoir (Sands)
Yes/No	Gas Hydrate Stability Zone (GHSZ)
Yes/No	Groundwater Feeder System
Yes/No	Sufficient Gas Flux (Any Source)

NGH concentration over a relatively short geological time. There has to be suitable host sediments in which the NGH can concentrate. Finally, there has to be a sufficient concentration of dissolved gas and existing physical chemical conditions immediately below the seafloor to provide a strong growth dynamic for NGH.

These elements may not have persisted in their present form for long in geological time, at least within the sediment section in which NGH may occur. NGH is very responsive to changes in near-seafloor environment conditions whereas conventional hydrocarbon deposits are not. Changes in ocean seafloor temperature, the local geothermal gradient, and the height of sea level stand, have no effect on conventional hydrocarbon deposits but exert a strong influence over the formation, concentration, and persistence of NGH. The geological age of oceanic sediment in which NGH concentrations may be found will be the youngest sediments deposited, as these will occupy the GHSZ unless erosion has occurred. Along continental margins where there is a high rate of sedimentation, these will tend to be Plio-Pleistocene in age.

However, sediment deposition is rarely evenly distributed along a continental margin, which is a reason why understanding the one-time courses of rivers and shelf sediment redistribution systems may be important to identifying those regions of a continental margin that are liable to have the most favorable host sediments in which NGH concentration could form. In a region as limited in scale as the northern Gulf of Mexico, which has the Mississippi River as its principal source of sediment, the depocenter has changed from the west to the east over time so that while the current depocenter Plio-Pleistocene sediments are prevalent in the west, NGH, or at least BSR, occurs in sediments of Miocene age (pers. observation, A.J).

Older sediments containing NGH may have slightly different properties than younger sediments at the same burial levels simply because dewatering of the sediments is partially a time, as well as an overburden compaction factor. Because the host sands are framework supported, they can be expected to have compacted relatively little with only the overlying marine sediment load. Sands in the GHSZ will have approximately the same porosity and permeability as they had following their initial compaction after deposition and burial. In contrast, finer grained sediments will continue to compact under increased sediment load well after sands have compacted to near their maximum extent. Where erosion of older sediments has taken place, bringing older sediments closer to the surface and into a GHSZ, the sands can be expected to provide hosts comparable with younger sediments in the GHSZ.

5.1 Sufficient Gas Source/Flux and the BSR

The most common gas component of NGH is methane, which appears to be primarily of biogenic origin. This has been confirmed in passive continental margins such as the Blake Ridge off the SE U.S., where deep thermogenic sources do not appear to have been tapped, and carbon isotope data indicates that biogenic methane dominates (Paull and Ussler 2001). Even in active margin areas, however, biogenic methane is much more common than thermogenic methane (Kastner 2001). Biogenic gas directly feeding GHSZ has been observed in drill holes (Wellsbury and Parkes 2003; Wellsbury et al. 2001; Wellsbury et al. 2000). Of the many drill holes into oceanic NGH, only a few have more than a few percent of thermogenic gas or traces of liquid hydrocarbons (Kvenvolden 1988). Where deeper hydrocarbon sources are tapped by deep faults, thermogenic gas may be a locally prominent component of the gas mixture along with traces of liquid petroleum. Such deep faulting is common in accretionary margins such as Cascadia (Trehu et al. 2004).

Natural gas liquids (NGL) of higher carbon number gases such as ethane, propane, and butane that might be expected to form under GHSZ pressure and temperature conditions have not been observed in association with petroleum-related thermogenic NGH. The NGL will not be present so long as there is water for them to react with. These gases have a stronger preference for forming NGH than methane, and complete sequestration of them in compound NGH in the presence of water is the rule. NGH is rarely associated with liquid hydrocarbons, even when the gas has a thermogenic source, although in some rare cases, where a very active conventional petroleum system leaks to the seabed, NGH and liquid petroleum may occur together, as has been observed locally in the Gulf of Mexico (Sassen et al. 2001).

Passive margins without deep faulting that could tap a thermogenic deep petroleum system, such as exist on the basin/continental margins of both the Amerasia and Eurasia Basins in the High Arctic, appear to be overwhelmingly sourced by biogenic methane. The NGH concentrations can be expected to be of high purity. This could be important to opening NGH exploration and production in the Arctic Ocean because the risk of pollution in the event of an accident that might release some gas is extremely low.

Apart from gas that can be observed naturally venting from the seafloor, the presence of bottom-simulating reflectors (BSRs) on seismic sections is a first order evidence for natural gas production and retention, but BSRs give little direct evidence about the potential for NGH concentrations. A BSR is useful only in the very early stages of exploration. The BSR is a reflection at a negative acoustic impedance contrast caused by free gas in the sediment below NGH in the GHSZ. Acoustic impedance is calculated as the product of compressional wave velocity times density; an acoustic reflection is produced at any interface where a contrast in acoustic impedance exists.

The presence of gas below the BSR markedly lowers the velocity and the gas NGH above increases the velocity somewhat compared to water-saturated sediments. A BSR does not mean that a gas column above water is present, especially in muddy sediments with low permeability where diffusion may be the primary mechanism of gas migration. Where an inclined permeable horizon crosses into a GHSZ, however, it is common to find gas pooled below NGH in pore space. Depending on the thickness of the permeable horizons, velocity analysis can be used to estimate both the NGH saturation and gas-water relationship (Max 1990;

Lee et al. 1996; Tinivella 1999; Frye 2008; Lee et al. 2009; Inks et al. 2009; Aung et al. 2011). Estimates of leakage at the seafloor combined with gas and NGH in-place will allow estimates of gas flux to be made. A first order of approximation for adequate gas flux, however, will be provided by the existence of the NGH itself. If gas flux were not high enough, no NGH would form.

Even in the lower portion of the GHSZ, however, NGH concentrations with no BSR have been observed (Paull et al. 1998). This appears to have occurred when pore water solutions in the sediments immediately below NGH-enriched strata have been undersaturated with respect to gas generation so that no free gas is present. However, in the observed case, the saturation was apparently high enough to provide a driving force for NGH crystallization once the solutions reached the GHSZ. NGH concentrations can also form well above the base of the GHSZ from solutions that were relatively undersaturated and in which the driving force for NGH crystallization was too low until they have migrated to shallower depths where saturation increases as a function of lower pressure and where temperature is colder. These higher-level NGH concentrations will be indistinguishable from deposits formed from solutions formed lower in the GHSZ because the conditions governing crystallization will be the same.

In lower-grade deposits that tend to be finer grained (muddier) and less well bed-differentiated, continuous BSRs often occur at approximately the location of the base of the GHSZ and may extend over large areas. BSRs, whose importance has been overemphasized in the past, often constitute first order features on seismic sections. These well-defined BSRs, such as are seen in the Blake Ridge area of the U.S. East Coast continental margin, are dramatic seismic features but are of limited exploration and economic value beyond identifying the region as a gas province. The NGH associated with these features often forms extremely large, low-grade deposits (Max et al. 2006) that have relatively small percentages of between 3 % and 8 % NGH in diffusely defined horizons throughout huge volumes of fairly uniform muddy sediments. These do not constitute primary exploration targets.

The total amount of biogenic gas at a typical sedimentary site results from a cumulative process that potentially may have gone on over a long period of time. Methane is created by bacteria in sediments within and below the GHSZ, but at temperatures lower than the 'oil window'. Where gas is formed below the GHSZ, it will tend to rise as bubbles and by diffusion. When it enters the GHSZ it can react with water to form NGH. As sedimentation goes on, new sediment will bury older strata. Heat flow will tend to remain the same, so the thermal gradient beneath the sea floor will remain constant, as the GHSZ will maintain a nearly constant thickness during sedimentation. The result is that isotherms will rise to accommodate the accumulating strata because any given isotherm will maintain a constant sub-bottom depth below the surface. Pressure, which is dependent on water depth, will not change much, so the warming of deeper strata as isotherms rise to maintain the thermal structure of the GHSZ causes an upward migration of the base of the GHSZ. The upward migration of the base of the GHSZ through the sediments will result in dissociation of previously formed NGH that had been at the base of the GHSZ. The released gas will migrate upward, nourishing formation of new NGH in the superjacent GHSZ. This combined process of formation of new biogenic gas and recycling of basal GHSZ NGH will have the effect of increasing the amount of NGH in deposits, and can go on as long as sedimentary accretion (including organic carbon) continues. Deep gas does not need to be sourced so long as the sediment/NGH accumulation system at depths less than the depth of the oil window functions appropriately.

5.2 Migration Pathways/Feeding the Thermodynamic Trap in the GHSZ

Methane in the NGH system would have been carried from depth in ground water toward the surface in both primary and secondary porosity. Exploration for NGH concentrations will literally 'follow the water' from a methane rich, subjacent groundwater source to a location within the GHSZ where spontaneous NGH crystallization will take place. Tracking methane and groundwater source within a relatively short distance below the GHSZ to NGH concentrations within the GHSZ is a fundamental aim of NGH exploration. The groundwater system within marine sediments on deep continental margins and continental slopes is the driver of the NGH system. In a passive margin the water drive is predominantly due to sediment compaction under gravity, while in an active margin, tectonics and fractures are likely to be more important than gravity alone. Of course, in continental margins with high sedimentation rates and high rates of input of organic carbon, much of the methane can be generated biogenically in the shallow sediments and recycled through the gas NGH system as the seafloor accretes.

Detailed studies of ancient migration systems that have concentrated conventional hydrocarbons can be important to the identification of conventional gas deposits. In addition, geological analysis of a basin's thermal and sedimentological attributes is part of petroleum system analysis because the conditions for hydrocarbon generation, and concentration, can begin early in the history of a deep basin. Subsequent source and migration systems may further charge existing hydrocarbon reservoirs or they may form entirely new ones, often in sediments not yet deposited when source beds and initial reservoirs were formed. Studies of ancient migration systems are not relevant to NGH exploration unless the system is still transporting gas and fluids toward the surface. NGH is dependent on existing gas supply. Inactive gas generation and migration systems are of no significance to existing NGH.

NGH concentrations that may contain enough gas to warrant extraction are similar to conventional gas concentration in two important ways. There have to be sources for the natural gas, and there have to be geological pathways through which the methane is transported, most commonly in pore water systems. But with existing NGH concentrations, the sources of the gas are much less important than that there is sufficient supply of dissolved methane in the groundwater now and in the immediate past. If the concentration of dissolved methane in the pore water is high enough, NGH will form and persist. Thus, one of the exploration tools vital to NGH exploration is an understanding of pore water movements and its chemical character as part of a groundwater supply system. Water sources must be tracked into the GHSZ and mapped with fracture systems and the orientation of geological strata to provide a predictive capability.

Conventional hydrocarbon reservoirs may be found at any depth; usually much deeper than NGH concentrations. Thus, conventional hydrocarbon deposits may be hot, often above the boiling point of water at surface pressure, which may necessitate very careful handling. In contrast, NGH concentrations are confined to the GHSZ, a zone that commonly parallels the seafloor and only extends to a limited and variable depth, and which will move upward with continued sediment deposition. The ambient temperatures Natural Gas Hydrate (NGH): Temperature of NGH deposits are also unlikely to be above 40 °C, and probably substantially below.

Oceanic NGH concentrations generally are not found in geological traps bounded by seals, as would be expected of conventional gas deposits. Although less permeable beds may bound or enclose the more porous horizons into which NGH-forming gas can migrate to form NGH, which might suggest a conventional trap, the trap itself is the NGH, which concentrates the gas and holds it in place. The sediment host or "reservoir" does not have to be different in any way from the same strata that might, for instance, extend below the GHSZ (Fig. 5.2). Where free gas underlies NGH in a contiguous horizon, it is consistent to suggest that the source of the gas was from mainly below, with upward migration feeding the NGH formation within the GHSZ as well as the ponded gas.

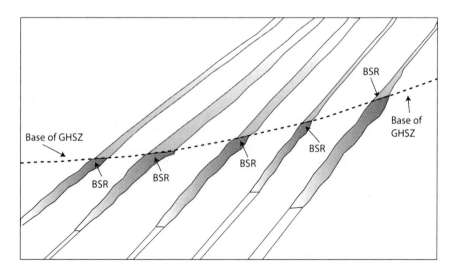

Fig. 5.2 Diagram of focused flow in permeable horizons showing the relationship of trapped gas below NGH at the base of GHSZ. *Note* non-continuous or 'string-of-pearls' BSR. Diagram based on Fig. 11 of Boswell et al. (2011). Extent of NGH up dip may be exaggerated. Angle of intersection of base GHSZ and permeable bedding is exaggerated. *Shading* indicates gas becoming less common down dip and NGH concentration more pronounced in lower part of GHSZ

While the porosity of host sands of the more permeable beds may comprise the reservoir of a conventional gas or oil deposit, NGH is unlikely to completely fill all available permeability. From a conventional gas viewpoint, the host sediment of a NGH concentration could be regarded as the 'reservoir', but an NGH concentration is different from a conventional gas deposit in that the NGH will commonly vary in its saturation of pore space. In a conventional gas deposit, gas pressure in the reservoir equilibrates, resulting in equal volumes of gas per pore volume. Therefore, in a NGH prospect, 'valuation' uses multiple cell analysis similar to that for more conventional mineral deposits.

Although Fig. 5.2 shows NGH forming at the base of the GHSZ, it may also occur at higher levels within the GHSZ. This can be a result of sudden gas infusion, for instance along faults, or formation of a secondary NGH zone where pore water of sufficient gas concentration reaches sufficiently cold temperatures in its ascent to allow NGH to form (Max et al. 2006). In any case, subjacent gas similar to that just below the base of the GHSZ is unlikely to be found in association with NGH higher in the GHSZ.

Potential NGH-forming groundwater solutions passing into the GHSZ are likely to begin crystallization in strata or in fracture zones that are no different in any way to their continuations or analogs at depth. Conventional hydrocarbon accumulations are not dependent upon a narrow set of physical-chemical parameters whereas NGH concentrations are. Existing or modern groundwater systems, which may have older geological antecedents of hundreds of thousands to millions of years, are critical to the existence of NGH concentrations. But under the right circumstances of high gas flux and natural refrigeration, NGH concentrations can develop relatively quickly.

As with the issue of gas sources, with which this aspect of the system is intertwined, the presence of substantial concentrations of natural gas indicates that migration pathways exist, as the local TOC within the existing GHSZ would rarely be high enough to provide all the natural gas held in the NGH. High volumes of NGH would indicate good operation of the gas flux/migration pathways system. The presence of thermogenic gases would be a clear indication that deep source gas has migrated into the GHSZ. The paucity of thermogenic gas in NGH analyzed to date indicates that deep sourced thermogenic gas is not a prominent source unless microbial activity is consuming heavier hydrocarbons to yield methane.

5.3 NGH High-Grade Reservoirs

Almost all conventional hydrocarbon reservoirs are hosted in sediment beds having relatively high porosity and permeability. Although NGH is known to form in fracture zones as well as sediment beds, Primary porosity zones rather than fracture zones are the primary NGH objective.

5.3.1 GHSZ Thickness

Geothermal gradients and seafloor temperature are used to determine the thickness of the GHSZ, but there is not enough heat flow or sediment thermal conductivity data for the Arctic Ocean Basin to allow geothermal gradients to be calculated on the scale of individual concentrations of possible economic value. There have been few measurements made since the early 1990s, and these have been along the Greenland-northern Canadian shelf and slope or in the Eurasia Basin. Davies and Davies (2010), for instance, show only about 40 measurements in the Arctic Ocean. By analogy with the GoM, where marine sediments may rest on a continental margin and oceanic crust as old as mid-Jurassic, a similar thickness of GHSZ can be anticipated above both subjacent sediments so long as the gas generating character of the sediments and the availability of migration paths are similar. A heat flow similar to the northern GoM is assumed for the Arctic for the same reasons as discussed by Max and Lowrie (1993). Because the temperatures of the seafloor in the Arctic may be as much as 10 °C cooler than that of the lowermost seawater in the GoM a considerably thicker GHSZ can be expected, at least in the Amerasia Basin.

There is very little reliable geothermal information from the Arctic Ocean. The relative youth of the oceanic crust in the Eurasia Basin may be associated with a relatively high heat flow that would tend to thin the GHSZ. The GHSZ of the Eurasia Basin is probably thinner near its northern margin along the Amundsen Basin (Kristoffersen and Mikkelsen 2004) and its underlying Basin where the heat flow is unexpectedly high, than along the Barents shelf—Nansen Basin margin where heat flow is more normal (Urlaub et al. 2010). But the area with highest heat flow appears to be in an abyssal region in which the Amundsen—Pole Abyssal Plain fill a basin. No high-grade NGH are likely in the thin sediments along the Lomonosov Ridge margin, which would have been sediment starved once spreading had opened the Eurasia Basin. The thicker sediment wedges along the northern Barents Sea margin and the northern North Atlantic margins offer promise for source and host sediments.

5.3.2 Suitable Sediment Hosts (Turbidite Sands)

NGH is developed in the pore spaces of unconsolidated sandy sediment near the margin of the deep submarine-fan turbidite system in the play SE of Tokyo Japan (Tsuji et al. 2004; Uchida et al. 2004; Noguchi et al. 2011; Egawa et al. 2013) and in continental slopes, as they are in the GoM (Boswell et al. 2011). NGH has also been identified in turbidite sands in the Ulleung Basin east of Korea (Lee 2011). They can be anticipated in the Arctic continental margins for the same reasons that they exist on other continental margins. Both have high sediment supply, including sands and coarse horizons, relatively quiescent depositional environments acting within the framework of sequence stratigraphy related to sea level rise and fall, and active crustal tectonics. The elements of these are described in many books, reports, and papers and will not be reviewed here. With some notable exceptions, sequence stratigraphy models generally concern sediment deposited primarily on the continental shelf during interglacial periods where it is subject to winnowing, with transport into deep water as sea level drops and/or the shelf rises owing to tectonism.

Departures from the basic sea level variation model can exist where there is strong longshore drift of sediment along a coast during an interglacial. If this is combined with the presence of a submarine canyon that cuts into the shelf (such as the Monterrey Canyon or the Rhone Canyon), the sediment will move down the canyon and out into deep water. Similarly, where a shelf is narrow, a major river can move a tremendous amount of sediment off the shelf, but that tends to be a temporary situation and can change as the depocenter for a delta changes. For example, the Mississippi River Delta, where the main channel is currently at the shelf edge, pours much of its sediment almost directly onto the continental slope. Other previous and possibly future main channels would deposit their sediment over 100 miles from the shelf edge, where it would be subject to considerable winnowing. Seismic data near the GoM outer shelf shows a number of channels from Wisconsin, Kansan, and Nebraskan lowstands that were filled during the interglacial periods.

Sediment deposition along the Arctic continental margins can be expected to have varied through time in a manner very similar to that of the northern GoM. At least three of the prolongations of the major Russian rivers flowing onto the continental shelf, the Ob, Yenesey, and Lena, could have carried sediment loads comparable to the Mississippi out to the edge of the continental shelf at lowstands. The Mackenzie River in Canada would also have carried large sediment loads during major erosional periods. The depocenter of the Mississippi River, which drained a large part of North America throughout the Tertiary and Quaternary, was sometimes focused to the west or east (as it is now). This resulted in thick sediment deposits being spread laterally along the continental shelf and slope. A similar tendency to distribute sediment broadly along a continental slope in the Arctic would likely have existed at the shelf-slope spill points of the Arctic river systems.

The age of the sediments within the GHSZ in the Northern GoM varies considerably. To the east (due south of the current Mississippi Delta), BSRs are developed in Miocene sediments. As the Pleistocene depocenter off the western Louisiana coast thins to the south with increasing water depth, the Pliocene seems to have the thickest section within the GHSZ. One of the oddest occurrences (and best logged) is the Chevron AC 818 well where drill holes through the unconformity encountered Oligocene NGH-bearing sands. Key factors in the thick GHSZ in the deepwater GoM are the very low geothermal gradients in the basins, high geothermal gradient over rooted salt structures, high-volume sediment deposition, and active local tectonics.

Very little is presently known about the turbidite systems of the deepwater Arctic region because of the lack of seismic data. An outline of sedimentation on

5.3 NGH High-Grade Reservoirs

the shelves and upper slopes (Levitan and Lavrushin 2009) indicate that sands commonly comprise 20 % of sediments with sand/silt up to 60 %. Surface sediments in the ocean basins and a few relatively short cores (Stein 2008) are not adequate to assess the likelihood of sand bodies in the GHSZ on the continental margins. A general review of the deeper shelf and slope areas with respect to river systems, glaciations, the continental shelf and sea level rise and fall (a basic element of sequence stratigraphy), however, provides an outline for exploration.

The geological and tectonic history of especially the Amerasia Basin is too sketchy for detailed basin-wide tectonic cycles to be understood well enough to define the tectonic megasequences composed of pre-rift, syn-rift, and post-rift phases (Jones and Underhill 2011). These control crustal subsidence and the depositional environments. The objective is to identify the continental margin areas that are most likely to have had active sand turbidite deposystems active over the last two million years or so. Like the northern GoM, where the Pleistocene is more than a mile thick over large areas of offshore Louisiana, the Arctic marine sediments, which are surrounded by landmasses subject to periods of high erosional glaciation, should have thick sections of Pleistocene, Pliocene, and Miocene age.

There are a number of elements that should contribute to a greater proportion of sands and coarser sediments on Arctic continental margins than are seen in the GoM. In contrast to the GoM, Arctic continental margins are relatively close to the erosional source areas. Much of the lower Mississippi River and its major tributaries have low slope gradients for over a thousand miles from the continental margin. This situation should increase the proportion of fines over sands and coarser sediment fractions, as sands will tend to be sidetracked and deposited along the course of the rivers, especially where natural levee systems develop. Because it was inferred for many years that there was little chance of deepwater sands in the GoM, energy companies did not explore the GoM deepwater. Exploration has now overturned this concept and has identified substantial deepwater sand hosts to hydrocarbon deposits. It is now known that there are substantial sand units in the continental slope sediments, with many of them within the GHSZ (Frye 2008).

Arctic shelves are very different from the GoM in having their erosional sediment source areas closer to the continental margin depocenters, especially during lowstands when sands deposited on the shelves as part of the glacial cyclicity could be mobilized. Where depocenters are close to the source of the erosion, proportionally more coarse-grained material can be expected to reside in the marine environment rather than along the terrestrial course of the rivers. The shelves have been emergent and subject to extreme erosional processes over extremely large areas for probably the entire glacial period in which seawater was sequestered in ice and began to be submerged with the onset of the interglacial. Sea levels are still rising. The most significant erosion and deposition toward the shelf edge and along the continental margin is likely associated with the transition from glacial to the interglacial stage when abundant water runoff, frost heaving, and sediment reworking across the continental shelves would have taken place as sea level rose from its glacial minima.

Even though there has been little tectonic activity along the Arctic Ocean margin that could act as a 'sand machine' and provide coarse sediments to the shelf and continental margin, there should be proportionally more coarse sediment available to be deposited in the continental deepwater sediments than along low-lying continental margins not having this history of extreme erosion and sediment deposition periodicity. Because of the influx of large amounts of sediment, a considerable proportion of which can be expected to be sand size or larger, and the high biological productivity and cold water that would favor burial of organic carbon, large concentrations of oceanic NGH are likely to be present in marine sediments on Arctic deep shelves and continental slopes.

There is a further reason why the Arctic deepwater sediments may contain proportionally more coarse-grained sediments than the GoM. Where ice cover or rivers reach near the continental margin, sediment derived by sub-ice erosion is deposited directly onto the continental slope. At lowstands, Arctic rivers would reach the sea (at about the present 120 m depth contour) very close to the continental margin. This can lead to the formation of marine hyperpycnal flows and substantial turbidites that are characteristically inversely graded (Mulder et al. 2003). Hyperpycnal flows could also be responsible for many of the sand turbidites in the GoM, at those times when the Mississippi mouth was very near the continental margin, as it is now.

The potential for winnowing on the northern GoM continental shelf is lower than on the Arctic shelves. Considerable sediment is incorporated within the GoM shelf as the sediment compacts. That is, much of the sediment, transported by the Mississippi River resides in the delta itself, rather than being carried to the slope. Although the present high-stand mouth of the Mississippi essentially dumps its sediment load directly onto the slope, this has not always been the case as the course of the lowermost Mississippi has changed over time. As sands would be expected to be deposited on the continental shelf, building the shelf seaward, the fine fraction reaching the slope would likely be enhanced.

In contrast, the emergent zone of the Arctic shelves, at least along the Eurasia margin, is much broader. Although the Barents shelf has deeper water that was ice covered during at least the last sea level low stand, all of the broad Siberian-Chukchi continental shelf that abuts Alaska would have been exposed during low stands. A broad shelf area allows greater winnowing to take place, which should further enhance the deposition of sands on Arctic continental slopes.

References

Aung TT, Noguchi S, Oikawa N, Kanno T, Tamaki M, Akishisa K (2011) Integrated facies modeling workflow for methane hydrate reservoir along the eastern Nankai Trough, Japan. In: Proceedings international petroleum technology conference, Bangkok, Thailand, p 8, 15–17 Nov 2011

Boswell R, Collett TS, Frye M, Shedd W, McConnell DR, Shelander D (2011) Subsurface gas hydrates in the northern Gulf of Mexico, marine and petroleum geology, p 21. doi:10.1016/j.marpetgeo.2011.10.003

References

Davies JH, Davies DR (2010) Earth's surface heat flux. Solid Earth 1:5–24. www.solid-earth.net/1/5/2010/

Egawa K, Furukawa T, Saeki T, Suzuki K, Narita H (2013) Three-dimensional palemorphologic reconstruction and turbidite distribution prediction revealing a Pleistocene confined basin system in the northeast Nankai Trough area. AAPG Bull 97(5):781–798. doi:10.1306/10161212014

Fiduk JC, Weimer P, Trudgill BD, Rowan MG, Gale PE, Phair RL, Korn BE, Roberts GR, Gafford WT, Lowe RS, Queffelec TA (1999) The Perdido Fold Belt, northwestern deep Gulf of Mexico, part 2: seismic stratigraphy and petroleum systems. Am Assoc Pet Geol Bull 83(4):578–612

Frye M (2008) Preliminary evaluation of in-place gas hydrate resources: Gulf of Mexico outer continental shelf. US department of the interior minerals management service resource evaluation division OCS report MMS 2008-0004, p 136

Inks TL, Lee MW, Agena WF, Taylor DJ, Collett TS, Zyrianova MV, Hunter RB (2009) Seismic prospecting for gas-hydrate and associated free-gas prospects in the Milne Point area of northern Alaska. In: Collett TS, Johnson AH, Knapp C, Boswell R (eds) Natural gas hydrates: energy resource potential and associated geologic hazards: American association of petroleum geologists memoir, vol 89, pp 555–583

Jones DW, Underhill JR (2011) Structural and stratigraphic evolution of the Connemara discovery, Northern Porcupine Basin: significance for basin development and petroleum prospectivity along the Irish Atlantic Margin. Pet Geosci 17:265–384

Kastner M (2001) Gas hydrates in convergent margins: formation, occurrence, geochemistry, and global significance. In: Paull CA, Dillon WP (eds) Natural gas hydrates occurrence, distributions, and detection, American geophysical union geophysical monograph, vol 124, pp 67–86

Kvenvolden KA (1988) Methane hydrate: a major reservoir of carbon in the shallow geosphere? Chem Geol 71:41–51

Kristoffersen Y, Mikkelsen N (eds) (2004) Scientific drilling in the Arctic Ocean and the site survey challenge: tectonic, paleoceanographic and climatic evolution of the polar basin. Jeodi Workshop, Copenhagen, Denmark, 13, 14 Jan 2003, Geological Survey of Greenland, pp 83

Lee S-R (2011) 2nd ulleung basin gas hydrate expedition (UBGH2): findings and implications, In: Fire in the Ice 11(1), U.S. Department of energy methane hydrate Newsletter, 6–9

Lee MW, Hutchinson DR, Collett TS, Dillon WP (1996) Seismic velocity structure at the gas hydrate reflector, offshore western Colombia, from full waveform inversion. J Geophys Res 99:4715–4734

Lee MW, Collett TS, Inks TL (2009) Seismic-attribute analysis for gas-hydrate and free-gas prospects on the North Slope of Alaska. In: Collett TS, Johnson AH, Knapp C, Boswell R (eds) Natural gas hydrates: energy resource potential and associated geologic hazards. American association of petroleum geologists memoir, vol 89, pp 541–554

Levitan MA, Lavrushin Yu A (2009) Sedimentation history in the Arctic Ocean and Subarctic Seas for the last 130 kya. Lecture notes in earth sciences series, Springer Dordrecht Heidelberg London New York, p 387. ISBN 978-3-642-00287-8. doi 10.1007/978-3-642-00288-5

Max MD (1990) Gas hydrate and acoustically laminated sediments: probable environmental cause of anomalously low acoustic-interaction bottom loss in deep ocean sediments. Naval Research Laboratory Report 9235, p 68

Max MD, Johnson AH (2011) Hydrate petroleum approach to natural gas hydrate exploration. In: Proceedings of the 7th international conference on gas hydrates (ICGH 2011), Edinburgh, Scotland, UK, CD, Paper 637, 12 pages, 17–21 July 2011

Max MD, Johnson AH (2013) Natural Gas Hydrate (NGH) Arctic Ocean potential prospects and resource base. OTC Paper 23798. (Digital) proceedings arctic technology conference, Houston, Texas, USA, 3–5 December 2012, pp 11

Max MD, Lowrie A (1993) Natural gas hydrates: Arctic and Nordic Sea potential. In: Vorren TO, Bergsager E, Dahl-Stamnes OA, Holter E, Johansen B, Lie E, Lund TB (eds) Arctic geology and petroleum potential, proceedings of the norwegian petroleum society conference, Tromso, Norway, 15–17 Aug 1990. Norwegian petroleum society (NPF), Special Publication 2. Elsevier, Amsterdam, pp 27–53

Max MD, Johnson A, Dillon WP (2006) Economic geology of natural gas hydrate. Springer, Berlin, Dordrecht, pp 341

Mulder T, Syvitski JPM, Migeon S, Faugeres J-C, Savoye B (2003) Mar Pet Geol 20:861–882. doi:10.1016/j.marpetgeo.2003.01.003

Noguchi S, Furukawa T, Aung TT, Oikawa N (2011) Reservoir architecture of methane hydrate bearing turbidite channels in the eastern Nankai Trough, Japan. In: Proceedings of the 7th international conference on gas hydrates (ICGH 2011), Edinburgh, Scotland, UK, p 9, 17–21 July 2011

Paull CK, Ussler W (2001) History and significance of gas sampling during DSDP and ODP drilling associated with gas hydrates. In: Paull CA, Dillon WP (eds) Natural gas hydrates occurrence, distributions, and detection, American geophysical union geophysical monograph vol 124, pp 53–65

Paull CK, Borowski WS, Rodriguez NM, ODP Leg 164 Shipboard Scientific Party (1998) Marine gas hydrate inventory: preliminary results of ODP Leg 164 and implications for gas venting and slumping associated with the Blake Ridge gas hydrate field. In: Henriet J-P, Mienert J (eds) Gas hydrates: relevance to world margin stability and climate change: geological society London special publication, vol 137, pp 153–160

Sassen R, Sweet ST, DeFreitas DA, Morelos JA, Milkov AV (2001) Gas hydrate and crude oil from the Mississippi fan foldbelt, downdip Gulf of Mexico salt basin: significance to petroleum system. Organic Geochemistry 32, 999–1008

Stein R (2008) Arctic Ocean sediments. Processes, proxies, and paleoenvironment. Developments in marine geology 2. Elsevier, Amsterdam, The Netherlands, p 592. ISBN: 978-0-444-52018-0

Tinivella U (1999) A method for estimating gas hydrate and free gas concentrations in marine sediments. Boll Geofisica Teorica Appl 40(1):19–30

Trehu AM, Long PE, Torres ME, Bormann G, Rack FR, Collett TS, Goldberg DS, Milkov AV, Riedel M, Schultheiss P, Bangs NL, Barr SR, Borowski WS, Claypool GE, Delwiche ME, Dickens GR, Gracia E, Guerin G, Holland M, Johnson JE, Lee Y-J, Liu C-S, Su X, Teichert B, Tomaru H, Vanneste M, Watanabe M, Weinberger JL (2004) Three-dimensional distribution of gas hydrate beneath southern Hydrate Ridge: constraints from ODP Leg 204. Earth Planet Sci Lett 222:845–862

Tsuji Y, Ishida H, Nakamizu M, Matsumoto R, Shimizu S (2004) Overview of the METI Nankai Trough Wells: a milestone in the evaluation of methane hydrate resources. Res Geol 54:3–10

Uchida T, Lu H, Tomaru H, The MITI Nankai Trough Shipboard Scientists (2004) Subsurface occurrence of natural gas hydrate in the Nankai Trough area: implication for gas hydrate concentration. Res Geol 54:35–44

Urlaub M, Schmidt-Aursch MC, Jokat W, Kaul N (2010) Gravity crustal models and heat flow measurements for the Eurasia Basin, Arctic Ocean. Mar Geophys Res 30:277–292. doi:10.1007/s11001-010-9093-x

Wellsbury P, Goodman K, Cragg BA, Parkes RJ (2000) The geomicrobiology of deep marine sediments from Blake Ridge containing methane hydrate (Sites 994, 995 and 997). Proc ODP Sci Results 164:379–391

Wellsbury P, Mather ID, Parkes RJ (2001) 19. subsampling RCB cores from the Western Woodlark basin (ODP Leg 180) for Microbiology. In: Hichon P, Taylor B, Klaus A (eds) Proceedings of the ocean drilling program, Scientific Results vol 180, pp 12

Wellsbury P, Parkes J (2003) Deep biosphere: source of methane for oceanic hydrate. In: Max MD (ed), Natural gas hydrate: in oceanic and permafrost environments, 2nd edn. Kluwer Academic Publishers (now Springer), London, Boston, Dordrecht, pp 91–104

Chapter 6
Path to NGH Commercialization

Abstract The first national NGH research program was initiated by U.S. Department of Energy Research Center (now NETL) in Morgantown, WV. This produced a body of work that generated considerable interest and confirmed that NGH could be a potential natural gas resource. Since then, considerable progress has been made in understanding the NGH genesis as part of a NGH petroleum system. Seismic exploration processing incorporating geotechnical effects of NGH formation has been developed sufficiently so that discoveries can now be brought to the level of a prospect. Japan established its national program in 1995 and has completed the world's first technical production test of oceanic NGH on the 40 TCF Nankai NGH deposit in accordance with a planned timeline during March 2013. Part of the Nankai deposit is scheduled for production in 2018, which is only 5 years from the first production test. This is a near-term development timeline consistent with conventional deepwater field development. Other NGH developments may also be of a more near-term nature than has been thought possible until very recently.

Keywords Unconventional gas • Nankai • Natural gas hydrate • Near-term • Energy • Seismic analysis • Prospect • NGH

The first national NGH research program was initiated by Rodney Malone at the U.S. Department of Energy research center (now National Energy Technology Laboratory) in Morgantown, West Virginia. This produced a body of work that stimulated others to see NGH as a potential resource that could have economic proportions rather than merely as a geochemical oddity. Japan established its national program in 1995. The United States established a gas NGH program by statute in 2000. India and China, prominent amongst a number of other countries, now have on-going NGH research focused on energy. Countries deficient in their own energy/gas resources, such as India and Japan have the greatest incentive to produce NGH while countries that have abundant gas resources, such as the United States and Russia, have little incentive.

A number of energy companies have low-level NGH programs in order to be early adopters. The European Union and Scandinavian countries' main support of NGH research is focused on the part that NGH may play in global climate change. As the Arctic warms there will be increased access to the Arctic for energy exploration. It is likely that the hydrocarbon exploration and other commercial aspects of the Arctic, such as fishing and water transport of goods will be of increased interest and value to the Arctic nations, Norway, Canada, the United States, and Greenland-Denmark as the ice cover thins and the ice margin retreats. Resolution of resource rights issues is already underway in an international framework (IBRU 2011).

NGH is one of the principal unconventional gas concentrations, and the only one that has not yet been commercially developed. Indigenous coalbed methane, shale gas, and tight gas have gone from being minor gas resources 20 years ago to being major contributors (~38 %) of the U.S. natural gas supply in 2010–2011, with promise of national energy independence based on a gas economy. 20 years ago there was virtually no unconventional gas production. Now unconventional gas contributes over 35 % of the production gas base of the United States and there is potential for these unconventional resources elsewhere in the world. In particular, shale gas, which as recently as 2006 had minimal production, today provides 25 % of the natural gas in the United States. And further ramp up is underway. It is likely that NGH development would show the same rapid ramp-up in production following initiation of production as have coalbed methane and shale gas because most of the technology required for exploration and production can be leveraged from the existing industrial base.

In our opinion, NGH will probably follow the development trajectory of other 'unconventional' gas resources. With respect to the dependability with which exploration and production can now be carried out in these resources, they can hardly be considered to be 'unconventional' in any real sense. They are just different gas plays from the deeply buried, pressurized gas and oil deposits that have formed world energy base for over a century. Once a play becomes commonly and economically produced, it is simply part of the resource base.

There is emerging agreement that sand reservoirs containing NGH are the primary exploration objectives, not only because they appear to host most of the high-grade NGH concentrations (Ruppel 2011), but the geotechnical performance of the sand during NGH conversion to its constituent gas and water is almost certainly going to be more predictable and trouble-free than fracture-fill reservoirs. Sands, which have many of the characteristics of conventional gas deposits, are on the verge of successful production. Fracture-filling NGH has many of the characteristics of conventional gas deposits, where they are confined to geological strata and to metaliferous economic mineral deposits where they are in either preferentially fractured strata or crosscutting fracture zones. In a sand, the orientation of the body in which the gas will flow and concentrate are more predictive from analysis of seismic data than from a fracture system analysis in which small fracture zone interconnectivity may be more difficult to evaluate.

In addition, when NGH converts to gas and water, the overall mechanical strength of the reservoir decreases. Because sands are framework-supported beds, they might be expected to undergo minor compaction but to not necessarily become unstable. Dispersed NGH is similar to low grade metaliferous mineral deposits (Max et al. 2006). But converting NGH in a muddy horizon may cause sediment mass movements and unpredictable gas movements. In the case of both fracture-filling and dispersed NGH deposits, extraction models based on conventional mining practices may prove applicable to producing the far larger volume of gas in NGH not concentrated in sands.

Permafrost hosted NGH sand reservoirs are available now for conversion and production. According to Makogon et al. (1972), the Messoyakha Field in western Siberia has been producing natural gas from NGH for a considerable period of time as a function of simply depressurizing the subjacent gas reservoir by extracting natural gas. The hydrocarbon production history of the Russian Messoyakha Field, located in the West Siberian Basin, has been used as evidence that NGH is an immediate source of natural gas that can be produced by conventional means. Reexamination of available geological, geochemical and hydrocarbon production data suggests, however, that NGHs may not have contributed to gas production in the Messoyakha Field. More field and laboratory studies are needed to assess the historical contribution of NGH production in the Messoyakha Field (Collett and Ginsburg 1998).

NGH is unique among the unconventional gas sources in that it occurs in a number of different environments and has more than one petroleum system, depends on existing gas generation, migration and physical situation, and is sensitive to changes in its environment. Whereas coalbed methane, tight gas, and shale gas may have some differences in their relative geological settings and energy density, the production model for each is essentially consistent across the range of their variations; NGH has a unique mode of occurrence that may allow different production models.

The three main expressions of NGH concentrations are in: (1) sands, (2) fracture-filling, and, (3) dispersed NGH. Thicker veins associated with faults and nodules tend to occur with dispersed NGH. We agree with Boswell and Collett (2011) that NGH-enriched sands constitute a clear analog to conventional sand-hosted gas reservoirs, and it is likely that production from sands will be well established before other expressions of NGH are considered for production. These high grade NGH deposits constitute only a small proportion of the total NGH resource base (Fig. 6.1) but may be almost equivalent to the identified conventional gas resource base. Production from NGH sands will probably closely follow existing industry practices for gas production, with the additional practices developed for NGH conversion (Max et al. 2006; Max and Johnson 2011a, b, c), gas separation/concentration, and extraction.

Max and Lowrie (1993) incorporated new bathymetric and sediment thickness data for the Arctic Ocean that was held by the U.S. Navy (NAVOCEANO) and used analog assumptions about geothermal gradients based on minimal Arctic data to estimate the thickness of GHSZs for natural gas. Although the presence of

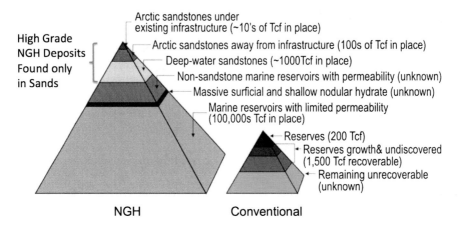

Fig. 6.1 NGH resource 'pyramid' and host strata types (*left*). Conventional reserves pyramid (*right*). After Boswell and Collett (2006). 'Arctic sandstones' (*topmost smallest sub-pyramid*) are onshore permafrost NGH-related. Economic potential decreases downward in both NGH and conventional 'pyramids'. Total NGH comparable with Max and Lowrie (1993)

higher density hydrocarbon gases would have the effect of increasing the thickness of GHSZ, data on gas composition is very sparse, and making assumptions is hazardous. It seems prudent to make estimates of GHSZ thickness using the assumption that the NGH is methane NGH, thus giving a minimum value for GHSZ thickness and providing a universal baseline. Long et al. (2008) used the GHSZ thickness map of Wood and Jung (2008), to calculate potential methane in NGH for Arctic continental shelf regions, although there is very little evidence that NGH underlies a substantial part of the continental shelf in which it could theoretically have formed during glacial sea level lowstand.

Max and Lowrie (1993) picked three methane-NGH likelihood zones in the Arctic continental margin (slope and rise) and abyssal basin areas sediments based on sediment thickness, likelihood of good source beds and regional gas province from known gas venting and an estimated percentage of NGH that could be present based on the relatively sparse existing drilling data, and heat flow and geothermal data to estimate NGH volumes. Their model was the Blake Ridge, which had drill hole locations picked from seismic surveys mainly carried out by the U.S. Geological Survey (Dillon and Paull 1983; Dillon and Popenoe 1988; Dillon et al. 1990). The Blake Ridge was the first major NGH location to be drilled (Paull et al. 1996, 1998). This allowed for refinement of seismic analysis techniques (Holbrook 2001) and led to model for NGH concentration in a diffuse flow framework. Identification of focused flow and high grade NGH concentrations (Max et al. 2006) was the next step in developing NGH exploration methodology.

In disseminated NGH, the NGH occasionally appears to occupy up to 8–10 % of the sediment mass. Wood and Jung (2008) produced a world map of methane-GHSZ thickness that included the Arctic Ocean using directly measured digital data and gridded data derived from Navy oceanographic chart used by Max and

Lowrie (1993). A Polar projection of this data was shown in Long et al. (2008) at a 2 min by 2 min (interpolated) scale. Although Wood and Jung's map shows considerably more apparent detail than Max and Lowrie (1993), especially in the troughs in the continental shelf for which Max and Lowrie had no consistent data, they caution that it should not be used directly for exploration because of uncertainty in the quality of the data input. Long et al. (2008) also show a map of predicted GHSZ thicknesses for the Arctic continental shelves. We note that where higher density natural gases such as ethane, propane, and butane from thermogenic sources would accompany natural gas into the GHSZ, the thickness of the GHSZ and the potential for natural gas held in NGH could be greater than for methane alone.

Previous estimates for NGH potential have relied on calculating percentage methane NGH predictions for the whole and parts of the predicted GHSZ for the Arctic as well as many other places. We regard this gross volume methodology for predicting NGH-in-place as now being outdated for the prediction of first-order potential economic deposits of NGH, although it still may be relevant for predicting NGH volumes for climate issues. Since the evaluation of the NGH resource potential in the Arctic Ocean by Max and Lowrie (1993), considerably more has become known about NGH and the manner that it is distributed in marine sediments, as well as about deepwater conventional deposits with which the gas NGH deposits may be aerially related in the northern Gulf of Mexico (GoM) (Halliburton 2008). Because this paper follows Max and Lowrie (1993) so directly, we have tried to not repeat descriptions, conclusions, or references where possible.

As an aid to guide exploration, we apply a NGH petroleum system analysis in which the various types of NGH accumulations are relate to associated sediments and depositional environments. This allows us to focus on the most valuable NGH deposits (Max and Johnson 2011a, b, c; Max et al. 2013). These new techniques are applied to the Arctic Basin and a methodology for evaluation of Arctic NGH resources, based on industry-standard techniques, is suggested by comparison with similar depositional environments and successful seismic analysis and drilling in the GoM. Although it is tempting to include the northernmost North Atlantic and especially the Labrador Sea in this discussion, we restrict this discussion to the Arctic Ocean alone.

As will be discussed elsewhere in this volume, very great progress has been made in understanding the NGH system and developing exploration tools that can bring discoveries to the level of a prospect. The world's first technical production test of oceanic NGH was carried out on the 40 TCF Nankai NGH deposit according to a planned timeline (Kurihara et al. 2011) during March 2013 by JOGMEC (2013). Part of the Nankai deposit is scheduled for production in 2018, which is only 5 years from the first production test. This is a near-term development timeline consistent with conventional deepwater field development. Commercial production of NGH off Japan is likely because natural gas produced from the Nankai NGH deposit should compete well with the rather high delivered price of liquefied natural gas (LNG) that has been in the $15–$18 MMcf range in the 2011–2013 time period. With improvement of the development cost of NGH exploration and production techniques, it is entirely possible that oceanic NGH may compete on a produced cost with other natural gas resources.

References

Boswell R, Collett TS (2006) The gas hydrates resource pyramid: fire in the ice, methane hydrate newsletter. US Department of Energy, Office of Fossil Energy, National Energy Technology Laboratory, (Fall Issue) 6(3), pp 5–7

Boswell R, Collett TS (2011) Current perspectives on gas hydrate resources. Energy Environ Sci 4:1206–1215. doi:10.1039/c0ee00203h

Collett TS, Ginsburg GD (1998) Gas hydrates in the Messoyakha gas field of the West Siberian Basin—a re-examination of the geologic evidence. Int J Offshore Polar Eng 8(1):22–29

Dillon WP, Paull CK (1983) Marine gas hydrates—II: geophysical evidence. In: Cox JL (ed) Natural gas hydrate: properties, occurrences and recovery. Butterworth, Boston, Massachussetts, pp 73–90

Dillon WP, Popenoe P (1988) The Blake Plateau basin and Carolina trough. In: Sheridan RE, Grow (eds) The Atlantic Continental Margin, vol 1–2. U.S. Geological Society of America. The Geology of North America

Dillon WP, Swift A, Booth JS (1990) Mapping sub-seafloor reservoirs of a greenhouse gas: methane clathrates. In: Proceeding of U.S. global climate change research program: marine geology research priorities, international symposium on marine positioning 9INSMAP 90, pp 15–19

Halliburton (2008) Deepwater gulf of Mexico 2000 to 2008 deepwater discoveries 1,000 Ft 4,999 Ft. PDF (1 image). www.halliburton.com

Holbrook WS (2001) Seismic studies of the Blake ridge: implications for hydrate distribution, methane expulsion, and free gas dynamics. In: Paull CA, Dillon WP (eds) Natural gas hydrates occurrence, distributions, and detection, vol 124. American Geophysical Union Geophysical Monograph, pp 235–256

IBRU (2011) Maritime jurisdiction and boundaries in the Arctic region. International Boundaries Research Unit, University of Durham. www.durham.ac.uk/ibru, p 3

JOGMEC (2013) News release. Gas production from methane hydrate layers confirmed. Mar 12 2013 www.jogmec.go.jp, p 3

Kurihara M, Ouchi H, Sato A, Yamamoto K, Noguchi S, Narita J, Nagao N, Masuda Y (2011) Prediction of performance of methane hydrate production tests in the eastern Nankai Trough. In: Proceedings of the 7th international conference on gas hydrates (ICGH 2011), Edinburgh, Scotland, United Kingdom, p 16

Long PE, Wurstner SK, Sullivan EC, Schaef HT, Bradley DJ (2008) Preliminary geospatial analysis of Arctic Ocean hydrocarbon resources. U.S. Department of Energy/Pacific Northwest National Laboratory PNNL-17922

Makogon YF, Trebin FA, Trofimuk AA, Tsarev VP, Cherskiy NV (1972) Detection of a pool of natural gas in a solid (hydrated gas) state (in English). Dokl Akad Nauk SSSR 196:203–206; Dokl Earth Sci 196:197–200

Max MD, Johnson AH (2011a) Diagenetic methane hydrate formation in permafrost: a new gas play? 2011. In: Proceedings, OTC Arctic technology conference, offshore technology conference, Houston, Texas, USA, 7–9 Feb 2011, p 7

Max MD, Johnson AH (2011b) Methane hydrate/clathrate conversion. In: Khan MR (ed) Clean hydrocarbon fuel conversion technology, woodhead publishing series in energy No. 19. Woodhead Publishing Ltd, Cambridge, UK, pp 413–434. ISBN 1 84569 727 8, ISBN-13: 978 1 84569 727 3

Max MD, Johnson AH (2011c) Hydrate petroleum approach to natural gas hydrate exploration, In: Proceedings of the 7th international conference on gas hydrates (ICGH 2011), Edinburgh, Scotland, UK, July 17–21, CD, Paper 637, p 12

References

Max MD, Lowrie A (1993) Natural gas hydrates: Arctic and Nordic Sea potential. In: Vorren TO, Bergsager E, Dahl-Stamnes ØA, Holter E, Johansen B, Lie E, Lund TB (Eds.) Proceedings of the norwegian petroleum society conference, Arctic geology and petroleum potential, Norwegian Petroleum Society (NPF), Tromsø, Norway, 15–17 Aug 1990, pp 27–53 (Special Publication 2 Elsevier, Amsterdam)

Max MD, Johnson A, Dillon WP (2006) Economic geology of natural gas hydrate. Springer, Berlin, Dordrecht, p 341

Max MD, Clifford SM, Johnson AH (2013) Hydrocarbon system analysis for methane hydrate exploration on Mars. In: Ambrose WA, Reilly JF II, Peters DC (eds) Energy resources for human settlement in the solar system and Earth's future in space, vol 101., AAPG MemoirAmerican Association of Petroleum Geologists, Tulsa, pp 99–114

Paull CK, Matsumoto R, Wallace P et al (1996) Proceedings of the ocean drilling program, Initial reports 164, Ocean drilling program, College Station, TX, p 623

Paull CK, Borowski WS, Rodriguez NM (1998) Marine gas hydrate inventory: preliminary results of ODP Leg 164 and implications for gas venting and slumping associated with the Blake Ridge gas hydrate field. In: Henriet J-P, Mienert J (eds) Gas hydrates: relevance to world margin stability and climate change, Geological Society London Special Publication 137, 153–160 (ODP Leg 164 Shipboard Scientific Party)

Ruppel C (2011) Methane hydrates and the future of natural gas, supplementary paper 2.4, the future of natural gas. MIT Energy Initiative. http://web.mit.edu/mitei/research/studies/documents/natural-gas-2011/Supplementary_Paper_SP_2_4_Hydrates.pdf

Wood WT, Jung W-Y (2008) Modeling the extent of earth's marine methane hydrate cryosphere. In: Proceedings of the 6th international conference on gas hydrates (ICGH 2008), Vancouver, British Columbia, Canada, 6–10 July 2008

Chapter 7
Gas Production from NGH: We Have All the Basic Tools

Abstract Conventional valuation methods of volumetric versus well performance and size-based distribution analogs are not available for NGH yet, and may not be directly applicable because NGH does not saturate its reservoir as evenly as pressurized free gas. Exploration for NGH concentrations will follow a process similar to that of conventional gas deposits, beginning with general characteristics and focusing on individual prospects. In the first phase, basin analysis is used to show the likelihood of the potential for NGH. The second phase uses remote characterization methods (primarily seismic) to narrow the search and actively identify potential NGH concentrations. The third phase uses both remote and direct (drilling) methods to characterize the economic nature of the concentration(s), including volumetric calculations similar to those made for other mineral deposits. Substantial information gathered from conventional hydrocarbon exploration can be used for NGH exploration, particularly with respect to the identification of a gas province within which NGH concentrations would be anticipated.

Keywords Unconventional gas • Conventional gas • Natural gas hydrate • Basin analysis • Gas hydrate stability zone • Exploration geophysics • Exploration drilling • NGH

The search for potentially commercial NGH deposits is essentially a search for turbidite sands within a GHSZ. The tools for conventional deepwater can be used, and NGH exploration can be combined with conventional hydrocarbon surveys. Exploration for NGH concentrations will follow a process similar to that of conventional petroleum system analysis, beginning with general characteristics and focusing on individual prospects (Max and Johnson 2013). In the first phase, basin analysis is conducted that includes depositional modeling and the identification of basic physical properties that often includes an application of knowledge from analogous regions. This phase can show the likelihood of the potential for NGH. The second phase uses remote characterization methods (primarily seismic) to narrow the search and actively identify potential NGH concentrations and volumetric estimates. The third phase uses both remote and direct (drilling) methods to characterize the economic nature of the concentration(s).

It is possible to illustrate the NGH development of a deposit with reference to two bodies of work in the GoM, both of which lead the way to setting the approach to NGH exploration in the Arctic Ocean region. The first of these involved numerical processing of seismic data in order to estimate sand and NGH values. The second involved more focused processing of seismic data and drilling that provided ground-truth for the predictions made on the basis of geophysical analysis (NETL 2011). Further work will take into account a number of issues that were not taken fully into account in the geophysical processing and allow for recalibration of some of the processing itself.

Frye (2008) was the product of a major team effort that involved the U.S. Minerals Management Service, the U.S. Geological Survey, the U.S. Department of Energy and a number of specialist contractors. Bill Shedd and Jesse Hunt carried out most of the workstation interpretation with over 100 3-D surveys being used. A large part of the statistical computer modeling was done by John Grace. The first drilling program carried out under the auspices of the Joint Industry Project (JIP) of the U.S. Department of Energy in 2006 had encountered NGH in fractures but no resource assessment was carried out. The 2009 JIP program targeted and encountered NGH-bearing sands in a number of different depositional environments identified in Frye (2008).

A key point of the 2009 program was to test the predictive model and published in a special issue of Marine and Petroleum Geology. The pre-drill estimates for NGH saturation of the three areas drilled were matched by the drilling results. In fact, the resource assessment showed that the geophysical predictions were conservative. It is not intended that the excellent figures in these publications should be reproduced here; it is only intended to draw attention to the processes and their application to NGH exploration. We are only at the beginning of the NGH development cycle. With improvements to this first integrated exploration effort (the Nankai Trough off Japan was initially drilled on the basis of BSRs alone, with essentially no NGH petroleum system approach) the NGH petroleum system methodology applied in these GoM studies provide a template for NGH exploration in the Arctic Ocean and elsewhere.

Conventional valuation methods of volumetric vs well performance and size-based distribution analogs are not available for NGH yet, and may not be directly applicable because NGH does not saturate its reservoir as evenly as pressurized free gas in a conventional reservoir. The methodology for oceanic NGH petroleum system analysis follows the general principles of Max et al. (2006) governing NGH nucleation and growth and the manner in which the mineralization is formed and concentrated.

7.1 Phase 1. Basin Analysis

1. For any region, charts and increasingly available digital data and maps derived from them provide the bathymetry and sediment distribution. In the Arctic, the primary frame of reference is Jakobsson et al. (2004, 2008). The amount of information available varies considerably from place to place. For instance, in the GoM, where decades of exploration and drilling has yielded immense datasets and publications, very detailed numerical analysis such as that undertaken by Frye (2008) can be accomplished. In the Arctic, however, data is much more sparse and studies of that type require new surveys.

    At the basic level, only enough information needs to be accessed to allow for the general geological framework and distribution of erosional-depositional environments to be understood. Although bathymetry and seafloor conditions are also important in conventional hydrocarbon exploration (where they are important for drilling strategy), bathymetry and seafloor temperature and pressure are critical to the thickness of the GHSZ below the seafloor. This information is available in the GoM, whereas it is not widely available in the Arctic. In the absence of detailed geothermal information, the base of the GHSZ may be established empirically from reflection seismic data. NGH exploration in the GoM points the way toward what must be accomplished in the Arctic.

2. Geological basin analysis in the GoM has been conducted in detail as part of conventional hydrocarbon exploration, including evaluation of stratigraphy and sequence stratigraphy to establish the possibility of beds having primary porosity or secondary porosity zones that could host NGH concentrations. This evaluation is comparable to the current practice of conventional early-stage reservoir analysis. The objective is to locate turbidite depositional systems that would bring sands into the basin. Considerable work bearing on this phase was part of conventional hydrocarbon exploration for deeper objectives, and thus was available and could be directly applied. Similar seismic exploration techniques are being used to localize deep water sand bodies in the North Atlantic. Dmitrieva et al. (2012) for instance, demonstrate their identification of sand-turbidite systems on Paleocene continental slopes and basins between the Shetlands and Norway. Where similar bodies of deep water sands occur within GHSZs, they are prospective for NGH concentrations.

3. Evidence for the presence of subjacent gas and groundwater access to and through the GHSZ was updated and compiled from the gas seeps and vents that are common in the GoM. Ideally, NGH-mineralizing solutions must be able to transit into the GHSZ to attain the greatest likelihood of high-grade NGH concentrations. This information is well known to the oceanographic and exploration community and required only analysis from the perspective of NGH.

4. Using geothermal gradient data and seafloor temperature, the base of the GHSZ was identified as a function of water depth and distance from the shore. The GHSZ thickness in the GoM is highly variable owing to the presence of salt diapirs that have a high thermal conductivity. This yields a thickness map that

resembles an irregular polka dot quilt. There have been no reports of large salt masses in the Arctic Ocean Basin. Hence, GHSZ thicknesses should be less variable, which is an aid to exploration. Although determining the top and base of the GHSZ is now a common deepwater practice for seafloor safety and drilling concerns, determining the base of the GHSZ is specific to the exploration for NGH.
5. The top of the GHSZ below the sulfate-natural gas transition zone was determined as part of the geotechnical study of the seafloor as part of the standard drilling safety requirements.

7.2 Phase 2. Potential Reservoir Localization

1. Finer-grained structural contour maps of sands with the acoustic physical properties associated with various degrees of NGH saturation were identified. In addition to the top and bottom of the NGH being identified on the scale of drilling targets, gas-rich zones below NGH in the sand were also identified (Frye et al. 2010, 2011; Boswell et al. 2011a, b). Seismic data was used to create digital structural contour maps on porous bed bases and tops. This procedure is similar to current practices being used to define potential conventional drilling targets. Existing computer analysis techniques can be directly applied using numerical estimates for NGH saturation of the sands using industry-standard workstations and one of a number of commercial software programs.
2. Isopach maps of strata having the potential to host the NGH were produced to guide drilling. This is also similar to current practices and defines 'reservoir' potential.
3. Other higher frequency, seismic reflection data were obtained. This was similar to industry practices for conventional hydrocarbons, especially for shallow hazard identification.
4. The more detailed geological host information with more precise velocity information was applied. This was the last step to preparation of a drilling plan.

7.3 Phase 3. Deposit Characterization and Valuation

Everything else, such as logging, sampling, reserve calculations, extraction modeling, among other direct sampling and measurements, follows as a result of drilling. However, the set of economic considerations are very different from valuing a conventional deposit. For instance, conventional deposits tend to be hydrostatic within a reservoir. That is, porosity may vary but whatever porosity there is will be fully filled with gas or petroleum in a mature deposit. Whether NGH forms in bulk, that is, unsupported and entirely within pore water, or affixed to a surface, it is a solid whose formation increases sediment strength and the bulk modulus. NGH values can be expected to vary within a mineralized bed in much the same way that low temperature strata-bound metaliferous mineral deposits do. Economic geological methods for estimating grade, reserves, and value are required to be used rather than conventional liquid and gas methods for the most accurate volumetric assessment.

References

Boswell R, Collett TS, Frye M, Shedd W, McConnell DR, Shelander D (2011a) Subsurface gas hydrates in the northern Gulf of Mexico. Mar Pet Geo 21. doi:10.1016/j.marpetgeo.2011.10.003

Boswell R, Frye M, Shelander D, Shedd W, McConnell D, Cook A (2011b) Architecture of gas-hydrate-bearing sands from Walker Ridge 313, Green Canyon 955, and Alaminos Canyon 21: Northern deepwater Gulf of Mexico. Mar Pet Geo 16. doi:10.1016/j.marpetgeo.2011.08.010

Dmitrieva E, Jackson C, A-L Huuse M, McCarthy A (2012) Paleocene deep-water depositional systems in the North Sea Basin: a 3D seismic and well data case study, offshore Norway. Pet Geosci 18:97–114. doi:10.1144/1354-079311-027

Frye M (2008) Preliminary evaluation of in-place gas hydrate resources: gulf of Mexico outer continental shelf. U.S. Department of the Interior Minerals Management Service Resource Evaluation Division OCS Report MMS 2008-0004, p 136

Frye M, Shedd W, Godfriaux P, Dufrene R, Collett T, Lee M, Boswell R, Jones E, McConnell D, Mrozewski S, Guerin G, Cook A (2010) Gulf of Mexico gas hydrate joint industry project leg II: results from the Alaminos Canyon 21 site. In: Proceedings of offshore technology conference, paper 20560, p 21

Frye M, Shedd W, Boswell R (2011) Gas hydrate resource potential in the Terrebonne Basin, Northern Gulf of Mexico. Mar Pet Geo 19. doi:10.1016/j.marpetgeo.2011.08.001

Jakobsson M, Mcnab R, Cherkis N, Shenke H-W (2004) The international map of the Arctic Ocean (IBCAO). Polar stereographic projection, scale 1:6,000,000. Research publication RP-2. U.S. National Physical Data Center, Boulder, Colorado 90305

Jakobsson M, Macnab R, Mayer L, Anderson R, Edwards M, Hatzky J, Schenke H-W, Johnson P (2008) An improved bathymetric portrayal of the Arctic Ocean: implications for ocean modeling and geological, geophysical and oceanographic analyses. Geophys Res Lett 35(5):L07602. doi:10.1029/2008GL033520

Max MD, Johnson A, Dillon WP (2006) Economic geology of natural gas hydrate. Springer, Berlin, Dordrecht, p 341

Max MD, Clifford SM, Johnson AH (2013) Hydrocarbon system analysis for methane hydrate exploration on Mars. In: Ambrose WA, Reilly JF II, Peters DC (eds) Energy resources for human settlement in the solar system and Earth's future in space, vol 101., AAPG MemoirAmerican Association of Petroleum Geologists, Tulsa, pp 99–114

NETL (2011) http://www.netl.doe.gov/technologies/oil-gas/FutureSupply/MethaneHydrates/JIPLegII-IR/

Chapter 8
What More Do We Need to Know?

Abstract Exploration for NGH in the Arctic deepwater is about at the same point as exploration for conventional deepwater hydrocarbons there. Because of the difficulty of access, little deepwater exploration and few seismic surveys exist in the deepwater in the central Arctic Ocean. Basic information, such as heat flow, is required to estimate GHSZ thickness. Subsurface data is required to establish the degree to which the sandy sediments exist in the continental slope sediments as these provide the best hosts for high grade NGH deposits. Subsurface data will also be used to calibrate general sedimentological analyses. Modern 3-D seismic surveys are required to allow full geotechnical analysis to be carried out, although 2-D survey can initially be used as a more general exploration tool. Moving more exploration activities to the seafloor will enhance year-round operations in the Arctic and accompany the move of production apparatus to the seafloor that is already taking place.

Keywords Unconventional • Seafloor • Natural gas hydrate • GHSZ • All-weather • Drilling • Exploration • NGH

NGH forms under very different conditions and has very different physical properties in its 'reservoir' than conventional gas deposits. Whereas a number of different petroleum systems can be established in a sedimentary basin over time, with the character of source beds, reservoirs, trap and seals, overburden and preservation all differing through time, NGH deposits worldwide have a single set of existing critical factors that controls their development. And that critical set of factors has to be active in the present or very recent past for NGH concentrations to form and persist.

Application of modern computer processing of seismic data in the northern GoM identified a considerable volume of sands with NGH. Frye (2008) shows many figures of processed images identifying these sands, which provided means of assessing drilling targets. The study was essentially a stochastic approach that was not designed to correlate directly with the drilling results, but did in fact predict not only where NGH would occur but also the nature of its development and rough % NGH/volume. A primary objective of the study was to map sands, and was very successful. Seismic analysis techniques for NGH developed by the NGH program of India have been summarized by Thakur and Rajput (2011). Further improvements in seismic analysis techniques based on laboratory analogs (Ye and Liu 2013; especially Lu et al. 2013) will allow further improvement to analysis.

NGH estimates, however, were made on the assumption that all of the gas was derived from immediately adjacent shales in order to maintain the same computational domains used in the numerical processing. This made Frye (2008) resource estimate very conservative, as gas influx from lower in the sediment pile was not included even though it is understood that considerable gas migrates into the NGH from subjacent sources. The JIP found that NGH occurrence was related to faulting and local pressure gradients, along with stratigraphy. This followed GoM exploration experience that suggested gas would preferentially fill downthrown fault blocks, which typically are at lower pressure than the adjacent upthrown blocks. The other thing the JIP did was to conduct detailed mapping of stratigraphy, with three very different types of deepwater sand bodies targeted.

Computer processing of seismic data (Frye 2008; Boswell et al. 2011) and drilling ground truth in the northern GoM has delivered the basis for workable industry-standard NGH exploration in which thick, highly saturated gas NGH-bearing sands were encountered in 4 of 7 wells drilled on the geophysical targets. The objective of the drilling was to document a range of gas/NGH systems, including both fracture-shale accumulations and low-saturation deposits in sands to ground-truth the seismic interpretation.

Industry standard hardware and software was used. Analytical techniques included attribute analysis, elastic inversion, and spectral decomposition. The seismic analysis tools were developed for identifying sandy units in a continental slope environment and for determining a wide variety of seismic anomalies that are directly applicable to identification of NGH concentrations using different assumption values for acoustic response based on the variable amounts of high Vp NGH. Although developed for conventional hydrocarbons, these techniques appear to be transferable to NGH exploration. The high level of success in relating acoustic prediction to drilling measurements appears to have validated the prospecting approach used in the selection of expedition targets. We may not need to know much more before NGH can be opened up as a gas resource.

8.1 Exploration Factors

Exploration for NGH in the Arctic deepwater is about at the same point as exploration for conventional deepwater hydrocarbons. Because of the hostile conditions and sea ice cover little deepwater exploration has been conducted and few seismic surveys exist in the deepwater areas. In shallow water Arctic shelves such as the generally ice-free Barents Sea, however, there is considerable seismic exploration data and production is slowly being pushed northward. This situation is different from that elsewhere in the world ocean where there is generally some relevant deepwater information available, and many areas where a great deal of seismic information configured for conventional hydrocarbon exploration exists. Normally, the first step in NGH exploration is to examine the available seismic data, taken for any reason. But there is little or no exploration data available in the Arctic Ocean deepwater to reexamine. In addition, there is virtually no drilling data and no more than minor sampling of surface sediments.

8.1.1 BSR Identification from Reflection Seismic Data

Because the BSR serves as first order evidence that some free gas exists in pore space, it can be used as a reliable indicator of gas flux. It also defines the base of the GHSZ. In addition, by estimating the acoustic velocities in the sediment column above the BSR, and assuming pure methane as the gas, geothermal gradients can be determined directly. This is especially useful where no heat flow and sediment thermal properties exist. Where old seismic data exists, this should all be reexamined. In areas that are difficult of access, even old single channel paper records can be useful in locating BSRs, even though most of the shallow data may have been processed out.

Because there is so little seismic information from the Arctic, it is recommended that seismic transits of opportunity using very simple setups with single, short arrays be carried out using minor ocean-going research platforms. This type of data is similar to transit data recovered, for instance, from aeromagnetic survey. If the arrays were long enough to allow velocity analyses to be carried out, this would provide valuable additional information, but in the first instance, a simple reflection seismic profile where none existed before would be valuable. These would not necessarily carry out parallel track surveys but would follow open water in ice-marginal and sea ice areas where enough leads existed to allow passage. These seismic transits of opportunity would provide valuable information at a reasonable cost and also provide tie lines for future surveys.

Where sea ice is less of a problem, wide-spaced parallel track surveys could identify sediment depocenters in which sands are likely to be concentrated. Chosen areas could be filled in with closer track data or even 3-D data where high possibilities exist for a high percentage of sand-rich sediments within a suspected GHSZ. Both seismic transits of opportunity and wide spaced surveys are a way of dramatically improving the seismic database quickly.

8.1.2 Heat Flow Data/Geothermal Gradients

Where detailed temperature/depth data is available from drilling (Inks et al. 2009), local variations in the base of the GHSZ can be contoured with confidence. The evidence that the base of GHSZ varies considerably even over small areas instructs us that generalizations of the depth to base GHSZ will remain no more than a good approximation. Determining the depth of the base GHSZ from seismic data and sediment thermal properties will remain the primary objective of determining GHSZ thickness as a basic exploration method.

Heat production from underlying continental crustal rock is in general higher than in oceanic crust, especially older oceanic crust that has cooled. In the Arctic, the continental margin sediments generally lie on continental crust, thinning as they lap out onto oceanic crust or crust of unknown affinity. The approximate junction between oceanic and continental crust in the Eurasia Basin is relatively well known but in the Eurasia Basin the nature of the crust in the deeper and abyssal regions is not well enough known to make good estimates of heat flow and geothermal gradient. Geochemical data can be used to estimate heat production in the lower crust and upper mantle. Geotherms are most sensitive to the upper crustal heat production so more accurate estimates of heat production are required to model their thermal state. By using compositionally corrected elevation and xenolith P–T estimates, it may be possible to make better estimates of upper crustal heat production (Hasterok and Chapman 2008; Hasterok et al. 2011).

In a NGH frontier area, and in the absence of seismically picked base of the GHSZ, being able to formulate a general thickness variation with depth is a first order exploration technique. For instance, Max and Lowrie (1993) made general predictions for NGH development using geothermal gradients analogous to those of similar crust elsewhere. When superimposed on sediment thickness, a GHSZ thickness could be defined. Generally, however, even the local heat flow variation along a shelf is not available from more than a few localities, in particular along the northern GoM, where this has been done as part of a NGH regional evaluation (Frye 2008).

The most prospective seismic data needs to be examined closely for evidence of BSR. Very prominent BSRs tend to occur in largely undifferentiated muddy sediments. These are not primary exploration targets because in these the NGH tends to be dispersed throughout large volumes. There are few permeable zones, such as sands, in which water and gas can flow easily. A 'string-of-pearls' BSR is a discontinuous one (Fig. 5.2). It occurs where porous beds and impervious beds are interbedded and where the succession dips through the base of the GHSZ. These are often quite difficult to pick, as the "pearl" occurs only where porous zones have a gas-NGH transition, and these zones may be some distance apart in the section. These discontinuous BSRs, however, are almost always associated with sedimentary hosts capable of holding high-grade class 1 or 2 NGH deposits (Table 3.1). The impermeable beds focus the groundwater and gas flow into the permeable beds, which are usually turbidite sands.

Source gas availability and groundwater velocity through the porous bed control NGH formation and subjacent gas trapping. If the system is open enough and enough porosity exists for water carrying dissolved gas to percolate through even after significant NGH has formed, conditions for gas accumulation at the base of the GHSZ may not exist. Assuming near-saturation conditions of gas in the groundwater, gas will accumulate below an almost fully NGH-saturated section with 'NGH-tightened' permeability.

8.1.3 Natural Gas Migration Path Analysis

The amount of NGH in high-grade deposits appears to be too large for the natural gas to have been generated from sediments found only within the existing GHSZ and immediately below it. In the Blake Ridge of the U.S. SE coast, for example, there is a huge amount of biogenic gas that is now resident in NGH that was generated above oceanic basement (Max et al. 2013, pp. 119–122 and Fig. 3.8). Thus, the natural gas in the zone and the fluids through which or along with which it migrated into the GHSZ probably has drawn on biogenic methane production through a considerable section of the subjacent sediments. Although it is not necessary to understand more than the 1 km or so of the subjacent natural gas-enriched feeder system, which may be the downward prolongation of the reservoir beds themselves below the base of the GSHZ, the movement of groundwater into the GHSZ is as important to development of high-grade NGH deposits as is the filling of a conventional reservoir with gas and petroleum. Better understanding of the groundwater that is carrying natural gas will lead to better NGH exploration, and to possible NGH recharge of the reservoir.

Application of groundwater techniques used to describe water movement in complex subsurface aquifer settings has yet to begin for the NGH petroleum system. Yet this is a well-understood field of study that could be applied to marine sediments. It is likely that analogous situations could be found that would allow indirect prediction of water movement.

Migration of hydrocarbons from source beds to reservoirs at 'critical moments' has been one of the basic elements of conventional petroleum analysis. Although modern seismic-numerical analysis techniques to increase the predictability of economic target picking (Sylta 2008) have little direct application to NGH because the critical moment is effectively now, some of the analytical techniques may be applied to seismic datasets in order to better describe the overall upward movement of fluids in marine sediments.

8.1.4 Exploration Drilling

There may be no better advice than to "drill, baby, drill" with respect to establishing a preliminary database of sediments within GHSZs on a wide variety of continental margins in order to define the full range of NGH drilling requirements. Only drilling will provide the directly measured data upon which innovative production modeling can proceed. Drilling to establish a NGH database in the Arctic, however, does not have to be as expensive as that for conventional hydrocarbons as a number of technical opportunities exist. We note that IODP (IODP 2011) presently is considering at least 7 proposals for drilling in the Arctic Ocean, which may include the use of smaller, mission specific vessels.

We would recommend that light drilling capability be employed to achieve cores and logging in the uppermost 1–1.5 km. This drilling would be inherently safe as the temperatures in the holes can be kept at ambient or lower (for any depth), NGH can be measured, sampled, and volumes estimated with no fear of gas release. Drilling risks mainly pertain to the possibility of conventional hydrocarbons occurring within the intended drilling location, but the shallow depth and a preliminary evaluation should minimize risk of a petroleum or gas release. However, drilling to more than 100 m should be done only along a control seismic line.

8.2 Production Factors

Because of the nature of the secondary recovery techniques that must be applied to convert NGH and produce the gas from it, there will be some issues that are unique for NGH or NGH in Polar Regions. But there are also a number of issues with which industry has experience and some existing technology that can be applied directly. There are also some aspects of production that will require substantial innovation to optimize.

8.2.1 Drilling in Preparation for Gas Production

Both light and heavy duty drilling capability is already being used in Polar Regions and existing drilling techniques are adequate to begin exploration, although technology opportunities exist to improve capability and cut costs. It is possible that long NGH Intersections will be required that can be provided through 'horizontal' and oriented drilling. These techniques are already being practiced and probably do not need to be much modified to be successfully applied to NGH production.

8.2.2 Thermodynamic Models for NGH Conversion

This is an area into which a considerable amount of work has produced conversion models. Moridis and Kowalsky (2006) and Moridis and Reagan (2007), amongst others have produced algorithm-based conversion predictions of energy transfer functions and likely costs for deposits in different pressure–temperature environments for oceanic class 2 NGH concentration in porous beds underlain by water production. These programs are now in common use. As they are applied to specific deposits, and as more is understood about *in situ* NGH conversion, improvements will be made. The existing programs, of which energy companies almost certainly have proprietary versions, are probably adequate to understand the major issues.

Seawater and marine sediment pore water freezes at about $-2°$ C, whereas pure water freezes at 0 °C. Where hypersaline brines formed as the reject water from the formation of sea ice sinks and resides on the seafloor, the freezing temperature is further depressed. In the uppermost part of the GHSZ, where temperatures are below 0 °C, experiments in dissociation in fine sands confirm positive gas yields (Wright et al. 1998), even though the fresh water produced by conversion may freeze. The production issues of NGH conversion involving freezing of water is a common issue with onshore permafrost NGH, in which some production testing has already been carried out.

8.2.3 Geotechnical Models

For safe production of the gas, the rate and position of NGH conversion to its constituent gas and water needs to be modeled so that the deposits, which are closer to the sediment surface and most likely in unconsolidated or only little consolidated sediments, do not dramatically weaken the structural integrity of the 'economic section'. The 'economic section' extends from the seafloor down to below any gas deposits that have been trapped by, and may be in hydraulic continuity with NGH in the GHSZ. It is important that conversion activities initiated to generate gas production do not cause any breeches in the integrity of the economic section that could lead to gas leakage or generation of sediment mass movements. These models may also be important where deeper geological traps of conventional hydrocarbons are the primary economic objective, and even in cases where no production from the NGH is planned.

Laboratory experiments (Winters et al. 2009) can only take us so far along the road toward a production model. Geological input is required to formulate a combined geotechnical/thermodynamic production model. For this reason, conversion tests in existing NGH deposits such as at the Mallik site and, more relevantly in the Nankai class 1 and 2 oceanic NGH deposits (3.1), are necessary steps. However, each deposit can be expected to have somewhat different thermal and physical properties that will require a dedicated production model. This follows the method for establishment of production models for conventional hydrocarbons.

Some issues that may affect the geotechnical stability are particular to NGH. For instance, because dissociation occurs at the boundary between NGH and its surrounding pore water media, some form of fracking may be advantageous for early production rate acceleration. Once NGH conversion begins, water–gas separation and both water and gas management issues that are new to gas production and may affect the geotechnical stability of the reservoir (although possibly similar to coalbed methane production) will need to be resolved.

8.2.4 Flow Assurance

The greatest flow assurance issue is the unwanted formation of new NGH that restricts or blocks flow. Flow assurance is a term usually applied to maintaining flow in hydrocarbon pipelines and flow lines. It is not the intent here to discuss inhibition or remediation of unwanted NGH, but only to note that flow assurance will be an important production consideration because the water and natural gas produced by conversion will each be saturated with dissolved fractions of the other and will spontaneously crystallize if conditions of NGH stability are allowed to reestablish. Flow assurance must be an inherent part of a production plan.

NGH is stable only at relatively cool temperatures, measured thus far at no more than 35 °C and more commonly below 25–30 °C (Max 2003). Gas produced from NGH is unlikely to be higher than 40 °C, even following heating that may be part of the conversion methodology. The temperature differential between the produced gas and the ambient temperature on the seafloor on which the wellhead and close-by transmission pipelines will be located will be no more than 40 °C. Therefore, the crystallization driving force and water vapor pressures in a NGH produced gas will be less conducive for unwanted NGH crystallization. Where existing conventional gas infrastructure is used to transmit the gas, it will already be insulated or have other provision for flow assurance. Only a small part of the existing flow assurance capability will be necessary to insure gas flow assurance. Where new infrastructure is used, it can be 'lighter' and amenable to innovative flow assurance measures of much lower cost than that used with conventional flow assurance.

8.2.5 Logistics and Infrastructure

Production of NGH in the Arctic will not have the deepwater infrastructure (deepwater to shore) of areas such as the GoM that was developed over a considerable period to service conventional hydrocarbon exploration and production. Apart from some infrastructure in the Barents Sea and in near shore areas off Siberia, there is essentially no infrastructure to share. Logistical issues, however, may not be as forbidding as they would at first appear. In the first instance, the Arctic Ocean is a relatively compact area in which deepwater exploration areas can be accessed with relatively short transits from any Arctic port. Year round ports are only found on the Norwegian and Russian Arctic coasts, although Nations without year round port access have declared strategic interests (White House 2013; USCG 2013). Sea ice is presently a problem, but it may become less of a concern as a result of climate change that presently is creating increasing open water and longer summers. Exploration can be expected to follow the retreating ice (sea ice cover maps are statistical, with ice cover shown when a certain proportion of a defined sub-area is over a certain percentage).

There are a very limited number of potential operational ports in the Arctic Ocean. Canada, Greenland, the United States, and Siberian Russia do not have suitable Arctic ports with suitable land communications. There are a number of ports along the northern rim of Norway and northern Russia, but Tromsø is probably the best significant year-round Arctic port, having good communications with industrial Europe, deep water, a settled community that can be expanded with due regard to environmental impact, and a dynamic Arctic culture. The Arctic port of Murmansk is available but other Russian ports east from Archangel are less well located.

Servicing exploration and drilling activities from the Pacific through the Bering Sea may actually be easier from the European ports even for the American waters of the Arctic Ocean. As of March 2012, the Shell Oil "Kulluk" drilling rig is scheduled to push off from the port of Seattle for the Beaufort and Chukchi seas (WP 2012). There currently is no other port in western North America (with the exception possibly of Vancouver) closer to the Arctic Ocean. Continued use of the Seattle-Vancouver port area will involve very long transits to the Arctic Ocean and must pass through the Aleutian Island Archipelago at the Unimak Pass between Unimak and Krenitzin Islands or further south to make the transit. Arctic operations from Alaska would have to be based from small ports in southern Alaska (which would still require a long transit and passage through the Unimak Pass) or from new ports constructed near the conventional production on the shallow water North Slope, which would be very expensive.

The development of a forward exploration and service port on Svalbard or on Russian islands along the northern margin of the Barents or Kara Seas is also possible. Mainland ports may be the most suitable for heavy operational support but coast guard and emergency rescue capability should be established near the deepwater exploration areas around the rim of the Arctic continental margin. Expanded exploration activity will require enhanced marine rescue and support capability.

8.3 NGH-Specific Technology Opportunities

The NGH zone of economic interest is much shallower than the deeper zones in which conventional hydrocarbons are commonly found. Exploration and extraction costs for NGH-optimized practices that are in the process of being developed may be significantly lower than those of conventional hydrocarbon exploration. The cost factors should be particularly divergent when NGH is compared with deep, high temperature, high-pressure conventional gas deposits. The comparatively accessibility of NGH deposits beneath the seafloor offers a number of opportunities for technology development and cost reduction over conventional deepwater hydrocarbon even though the requirement to dissociate the NGH into its components of gas and water requires additional processes.

Because of the special physical circumstances and properties of NGH and the highly reversible chemical reaction through which it forms and converts back to its constituent gas and water, its unique presence in thermodynamic rather than geological traps and in reservoir hosts having very similar drilling conditions within 1.3 km of semi-consolidated sediments worldwide, significant opportunities exist for development and implementation of NGH-specific exploration, drilling, and production technology. The generally soft semi-consolidated sediment containing the NGH drilling targets is essentially similar to semi-consolidated sediments worldwide. This allows less expensive and less costly approaches to be used. Smaller and more lightweight seafloor completion apparatus including much smaller blowout preventers and other links to riser pipe can be used.

Being able to control pressure within the payzone and production apparatus is a fundamental field where innovation is possible for NGH-specific technology and methodology. The pressures within a NGH deposit are the formation pressures. In contrast to conventional gas deposits that contain high-pressure gas that may depressurize as it is produced, NGH must first be converted (Max and Johnson 2011). Because conversion is a controllable process, the pressure within a deposit can be controlled. In fact, if depressurization is used* as the conversion methodology as it was in the recent Nankai test (JOGMEC 2013), the pressure in the payzone will be below formation pressures.

In addition, because the reservoir and production pressures of the gas will be relatively low by industry standards, less robust gas transmission processing facilities and pipelines will be required. This will reduce the overall cost of exploiting the NGH resource while increasing the commerciality of the resource. In addition, the relative chemical purity and low temperature of the converted gas, and the controllable gas pressures within the reservoir and collector systems also allow materials to be used that allow for additional new technology to be applied to NGH production.

The cost of producing natural gas from NGH is commonly calculated as a sum of the cost of NGH conversion from its stable form in its reservoir and current industry costs of conventional drilling and production activities. According to Collett (2010), "For both arctic and marine NGH-bearing sand reservoirs, there are no apparent technical roadblocks to resource extraction; the remaining resource issues deal mostly with the economics of NGH extraction."

By implementing a highly robotized (Robotics VO 2013) NGH-specific set of technologies and methodologies, to which the NGH specific costs are added, a much lower overall cost may be achieved, and this will have a strong impact upon the perception of commerciality of the NGH resource. It may be possible to keep capital expenditure and operating costs well below those currently envisaged by the current conventional model of pricing of NGH commerciality. By implementing new methodologies and technologies for maintaining high production rates, safety, and reliability, NGH may prove to have a cost profile that will render it highly competitive, even taking into account conversion costs. Low internal CAPEX costs will provide a lower base for further internal CAPEX reduction not available to conventional hydrocarbon exploration and production.

8.3.1 Moving to the Seafloor

Carrying out the majority of drilling and production operations from the seabed allows all operations to take place in almost isothermal conditions year round, well within the proven engineering limits of all existing deepwater materials. Seafloor temperatures in the world's open ocean are slightly above 0 °C and the temperatures that will be encountered during all drilling and production operations will be no more than about 50 °C. This temperature range means that proven materials can be used. In addition, storms such as occur in many regions that not only may cause surface drilling and other operations to cease do no occur on the seabed, except rarely in a few places where deep transient currents of 1–2.5 knots are thought to be restricted to the axes of very cold bottom water flows and along the margins of abyssal basins (Seibold and Berger 1996). These are regions in which NGH operations are not likely to take place. In any case, oceanographic and seafloor survey, in addition to sub-seafloor survey for resource assessment will be made and the site and equipment prepared accordingly.

Seabed operations thus are particularly practical in regions such as the Arctic Ocean and the Southern Ocean around the Antarctic continent, and in other regions where very cold weather is common and sea conditions are especially hostile. The essentially benign environment of the seafloor is very different from drilling operations from a surface vessel that will be exposed to the harsh conditions and low temperature of the Polar Regions. For instance, even if the sea ice thins and sea ice cover seasonally diminishes to the point where drilling can take place virtually anywhere in the Arctic Ocean, very cold temperatures will still occur throughout the Arctic in winter and a good part of the rest of the year. Existing materials may need substantial improvement to work in the Polar Regions. For instance, steel alloys that are proven to maintain their physical integrity at -40 °C can fail at -60 °C.

Even though the Arctic Ocean and Bering Sea may become essentially ice free for part of the summer, sea ice will still be present for most of the year. Therefore surface ships will still have to have ice strengthening and probably ice breaking support. Diminished sea ice in the summer may in fact increase hazards because drifting ice becomes more mobile and difficult to predict.

8.3.2 Drilling and Logging

Oceanic NGH occurs in similar marine sediment on continental margins worldwide. These will have relatively benign shallow drilling conditions so that much lighter and less expensive drilling apparatus and seafloor production systems may be utilized. NGH does not require the same heavy-duty drill ships and rigs because the stable nature of NGH in its reservoir is dramatically different from naturally dynamic conventional hydrocarbons. Full risers, for instance, may be unnecessary during the exploration phase because unless gas below the GHSZ is drilled into, there will be little or no naturally occurring gas to leak. The nature of NGH-specific drilling raises the possibility that not only can much smaller, lighter, and less expensive ship-borne drilling be carried out but also advanced seafloor drilling capability may be applied. For instance, coiled tubing or pipe cartridge drilling units could be established on the seafloor and used with LWD logging to recover data. Drill bits can also be optimized for sediment and NGH-sediment mixtures.

8.3.3 Undersea Processing and Completions

Communications and autonomous vehicle technology have reached the point where seafloor completion and primary processing equipment is increasingly being placed on the seafloor (Perry 2013). Continuing this trend would only involve acceleration of an existing trend that is being developed mainly for conventional hydrocarbon exploration and production, and is presently being applied for the anticipated production at Nankai, Japan. The engineering of systems to meet the less rigorous conditions of NGH production (chemically neutral gas and water at ambient temperatures rather than often gas, petroleum, and water at high temperatures with potentially dangerous chemical components) is already underway. Undersea completion and processing would be a great advantage in the high Arctic where winter sea conditions can be expected to remain very harsh. Undersea capability is the best solution to year round operation as well as conferring considerable overall risk reduction.

8.3.4 NGH: Specific Vessels/Seismic Survey

Academic-based marine geophysical research, however, has exploited the recent reduction of summer ice and is beginning to reveal geological details. The tantalizing new data, the warming of the Arctic responsible for reducing sea ice cover and disrupting the environment and indigenous inhabitants (human and otherwise) (Cooper 2005), and the economic need for enlarged energy resources, has initiated a political imperative to understand and exploit the energy resources of the high Arctic (CRS 2011).

NGH exploration has largely been carried out with ships and survey equipment designed for conventional hydrocarbon exploration, although recently, and especially in the Arctic, smaller vessels have been used for surveying and drilling. Advances in reducing the size, weight and cost of equipment, as well as enhanced computer power for the acquisition and processing of seismic and other data, have been serendipitously developed at just the time when NGH exploration and possibly production, presents less rigorous demands on equipment.

8.3 NGH-Specific Technology Opportunities

Seismic data for NGH exploration is only required from the upper 1 or 2 km of the marine sediment. For first order recognition of seafloor geology and evidence of gas flux, very simple and inexpensive seismic surveys can be carried out. For areas in which NGH deposits are anticipated, it is recommended dual surveys should be carried out. Deeper conventional hydrocarbon exploration can be combined with that for NGH. This is not so much a matter of using different equipment but of signal processing. Data can be split into two streams, each of which can be processed differently for the different depth regions in which they occur. This will involve treating the conventional data stream exactly as it is now and the NGH data stream according to processing suitable for delineating specific shallower NGH drilling targets (Frye et al. 2011).

8.4 Will NGH Deposits Be Commercially Competitive with Conventional Gas?

Because the figures for NGH are so large, it is possible that very large NGH deposits may be found. It is important to be able to relate NGH deposits to conventional gas. Giant oil fields are defined as having greater than 500 million barrels of oil or oil equivalent gas (in BTU equivalents. 1 BBL of oil = 6,000 ft^3 methane), so that a giant gas field with no liquids would be 3 Tcf, although some use 3.5 Tcf as the cutoff. A supergiant field has been variously defined as greater than 1, 5, or 10 billion barrels. Part of the difference lies in whether the figure refers to in-place or recoverable hydrocarbons. The most commonly used base line for a supergiant field (in our experience) is 5 billion barrels, or 30 Tcf gas.

The most well known NGH deposit is the Nankai accumulation to the SE of Tokyo, offshore Japan. Estimates of gas-in-place are divided into about 20 Tcf in NGH 'concentrated zones' and about 20 Tcf in "other than NGH concentrated zones" for a total of in excess of 40 Tcf gas-in-place (Fujii et al. 2008). This would place the Nankai NGH deposit firmly in the supergiant category. Even if only the gas in the 'concentrated' NGH were to be considered for conversion, the deposit would be equivalent to a very large giant gas field.

The only NGH deposit in US waters that has received an industry-standard workup is in the GoM. Early in 2009, following the workstation processing of GoM seismic data for NGH (Frye 2008), prospecting for NGH-bearing sands using traditional hydrocarbon exploration approaches were tailored for NGH. JIP Leg II tested this approach in the spring of 2009 by drilling seven wells at three sites. Of these wells, six encountered NGH in sand reservoirs consistent with pre-drill estimates (Boswell et al. 2011).

The Walker Ridge play would be considered potentially commercial with 740 Bcf of conventional gas, depending on proximity to pipelines, facility costs, and anticipated gas price during the first few years of production, although in the current U.S. gas market, no companies are pursuing deepwater gas prospects. On a BTU basis, 1 million barrels equals about 6 BCF of natural gas. But on a recent price basis, 1 million barrels is worth about 25–30 BCF gas. At this ratio, Walker Ridge would be equivalent to about 25 million barrel of oil on price comparison, although on a BTU basis it has the equivalent of 123 million barrels of oil.

The current oil/gas pricing ratio is probably not sustainable, and may decline to a ratio of 1 million barrels of oil to 15–20 BCF gas. At a very conservative 1 million BBL/15 BCF gas ratio, Walker Ridge's 740 BCF of conventional gas would be equal to about 50 million barrels, which would be still be uneconomic unless there were facilities close by to tie back to with subsea completions. The other factors affecting commercial viability are gas-in-place versus recoverable, certainty of reserve figures, and flow rates. NGH has the added factor of an unproven operating expense for conversion of NGH (Max and Johnson 2011), which will negatively affect the relationship between technically and commercially recoverability of NGH resources. The operating expense for producing conventional gas tends to be predictable and relatively low.

Gas and oil prices markets are very volatile and liable to remain so for the near-term. Increased demand, international uncertainties of secure transport, inflation driven by debt, and speculation will probably keep hydrocarbon prices on an upward trend over time. Until the impact of more widespread unconventional gas/oil production begins to rationalize supply that can be factored into the financial and gas markets world wide, gas price stability and predictable margins between wellhead and wholesaler/retailer prices will remain subject to sudden swings. With the development of more gas transport capability, an international base gas price and market similar to that which has existed for oil since WWII may be developing.

The simple answer to the cost competiveness of gas production of NGH is that commercial production is technically feasible but remains unproven, for the moment. In areas where weather is benign, and where potential production from deepwater sedimentary systems is relatively near shore and near markets, there is no underlying reason why NGH should not be competitive with gas imports. The degree to which remoteness, market and infrastructure availability, and logistics (including weather) increase costs can only be estimated in a very general way at present. But we are confident that the hydrocarbon industry, which has showed itself capable of rising to successfully meet challenges of increasing difficulty that were successively regarded by many as being irresolvable, also will innovate successfully in the field of NGH exploration and production.

References

Boswell R, Collett TS, Frye M, Shedd W, McConnell DR, Shelander D (2011).*Subsurface gas hydrates in the northern Gulf of Mexico. Mar Pet Geo 34:21. doi:10.1016/j.marpetgeo.2011.10.003

Collett T (2010) Petroleum systems based gas hydrate prospecting in the Gulf of Mexico. In: Proceedings of the geology of unconventional gas plays, Geological Society, London, England, 5–6 Oct 2010, pp 78–79

References

Cooper LW (2005) Proceedings of a workshop on facilitating U.S.—Russian environmental change research in the Russian Arctic, St. Thomas, Virgin Islands, 11–16 June 2005, p 71. (http://arctic.cbl.umces.edu/RAISE/index.html)

CRS (2011) Changes in the Arctic: background and issues for congress. Congressional Research service 7-5700, R41153, p 82. (www.crs.gov)

Frye M (2008) Preliminary evaluation of in-place gas hydrate resources: Gulf of Mexico Outer Continental Shelf. U.S. Department of the interior minerals management service resource evaluation division OCS report MMS 2008-0004, p 136

Frye M, Shedd W, Boswell R (2011) Gas hydrate resource potential in the Terrebonne Basin, Northern Gulf of Mexico. Mar Pet Geo 34:19. doi:10.1016/j.marpetgeo.2011.08.001

Fujii T, Saeiki T, Kobayashi T, Inamori T, Hayashi M, Takano O, Takayama T, Kawasaki T, Nagakubo S, Nakamizu M, Yokoi K (2008) Resource assessment of methane hydrate in the Nankai Trough. In: Proceedings of offshore technology conference, Paper 19310, Houston, TX, Japan, p 6. (http://www.mh21japan.gr.jp/english/mh21-1/2-2/)

Hasterok D, Chapman DS (2008) Global heat flow: a new database, a new approach. EOS transactions American geophysical union, fall meeting supplement 89, T21C-1985

Hasterok D, Chapman DS, Davis EE (2011) Oceanic heat flow: implications for global heat loss. Earth Planet Sci Lett 311:386–395. doi:10.1016/j.epsl.2011.09.044

Inks TL, Lee MW, Agena WF, Taylor DJ, Collett TS, Zyrianova MV, Hunter RB (2009) Seismic prospecting for gas-hydrate and associated free-gas prospects in the Milne Point area of northern Alaska. In: Collett TS, Johnson AH, Knapp CR, Boswell R (eds) Natural gas hydrates—energy resource potential and associated geologic hazards, vol 89. Tulsa, American Association of Petroleum Geologists Memoir, pp 555–583

IODP (2011) Scientific drilling in the Arctic Ocean, p 19

JOGMEC (2013) News release. Gas production from methane hydrate layers confirmed. March 12, p 3. <www.jogmec.go.jp>

Lu W, Wang F, Wang M (2013) Natural gas hydrates Experimental simulation of hydrate accumulation and dispersion in pore fluids. In: Ye Y, Liu C (eds) Natural gas hydrates experimental techniques and their applications. Springer, Berlin, pp 217–237 (Springer Geophysics Series XII). ISBN 978-3-642-31101-7

Max MD (ed) (2003) Natural Gas Hydrate: in oceanic and permafrost environments, 2nd edn. Kluwer Academic Publishers (now Springer), London, Boston, Dordrecht 422

Max MD, Johnson AH (2011) Methane hydrate/clathrate conversion. In: Khan MR (ed) Clean hydrocarbon fuel conversion technology, Woodhead Publishing Series in Energy No. 19. Woodhead Publishing Ltd. Cambridge, U.K. ISBN 1 84569 727 8, ISBN-13: 978 1 84569 727 3, pp 413–434

Max MD, Lowrie A (1993) Natural gas hydrates: Arctic and Nordic Sea potential. In: Vorren TO, Bergsager E, Dahl-Stamnes ØA, Holter E, Johansen B, Lie E, Lund TB (Eds.) Proceedings of the norwegian petroleum society conference, Arctic geology and petroleum potential, Norwegian Petroleum Society (NPF), Tromsø, Norway, 15–17 Aug 1990, pp 27–53 (Special Publication 2 Elsevier, Amsterdam)

Max MD, Clifford SM, Johnson AH (2013) Hydrocarbon system analysis for methane hydrate exploration on Mars. In: Ambrose WA, Reilly JF II, Peters DC (eds) Energy resources for human settlement in the solar system and Earth's future in space, vol 101., AAPG MemoirAmerican Association of Petroleum Geologists, Tulsa, pp 99–114

Moridis GJ, Kowalsky M (2006) Gas production from unconfined class 2 oceanic hydrate accumulations. In: Max MD (ed) Natural gas hydrate: in oceanic and permafrost environments, 2nd edn. Kluwer Academic Publishers (now Springer), London, Boston, Dordrecht, pp 249–266

Moridis GJ, Reagan MT (2007) Gas production from oceanic class 2 hydrate accumulations. In: Proceedings of the offshore technology conference 30 April–4 May, Houston, TX, p 33

Perry R (2013) Subsea processing technologies are coming of age. Offshore, p 4. <http://www.offshore-mag.com/content/os/en/articles/print/volume-72/issue-4/subsea/subsea-processing-technologies-are-coming-of-age.html>

Robotics VO (2013) A roadmap for U.S. robotics from internet to robotics (2013 Ed). Robotics Caucus Advisory Committee of the U.S. Congress, p 129. <www.robotics-vo.us/node/332>

Seibold E, Berger WH (1996) The Sea Floor An Introduction to Marine Geology (3rd Ed). Springer, Berlin, Heidelberg, New York. ISBN 3-540-60191-0, 337

Sylta Ø (2008) Analysing exploration uncertainties by tight integration of seismic and hydrocarbon migration modeling. Pet Geosci 14:281–289

Thakur NK, Rajput S (2011) Exploration of gas hydrates geophysical techniques. Springer, Berlin, p 281 (Springer Geophysics). ISBN 978-3-642-14234-5X

USCG (2013) The U.S. coast guard's vision for operating in the Arctic Region. Published by the United States Coast Guard. CG-DCO-X, Department of Homeland Security, p 47

White House (2013) National strategy for the Arctic Region. Issued by the White House in the name of the President of the United States who has signed the document, p 11

Winters WJ, Waite WF, Mason DH (2009) Effects of methane hydrate on the physical properties of sediments. In: Collett T, Johnson A, Knapp C, Boswell R (eds) Natural gas hydrates—Energy resource potential and associated geologic hazards, vol 89. Tulsa, American Association of Petroleum Geologists Memoir, pp 714–722

WP (2012) Shell rig prepares to drill exploratory wells in Arctic. Washington Post (attributed to the Los Angeles Times), p A2, Sunday March 12

Wright JF, Chuvillin EM, Dallimore SR, Yakushev VS, Nixon FM (1998) Methane hydrate formation and dissociation in fine sands at temperatures near 0°C. In: Proceedings of 7th international conference (proceedings), Yellowknife (Canada), Collection Nordicana # 55, pp 1147–1153

Ye Y, Liu C (eds) (2013) Natural gas hydrates experimental techniques and their applications. Springer, Berlin, p 402. ISBN 978-3-642-31101-7, p 402

Chapter 9
NGH Likelihood in the Arctic Ocean

Abstract We identify three different types or 'plays' of oceanic NGH in the Arctic deepwater. (1) The main resource potential is probably in continental slope sediments. The resource host here would be the same type of turbidite sand deposits in which conventional hydrocarbons are found at greater depths in these sedimentary prisms. Their genesis is well understood. (2) Because of the very cold seafloor water, the top of NGH stability may be initiated at relatively shallow water depths. NGH may also occur in Intra-shelf troughs, depressions and along deep shelf margins bathymetrically above a slope break that denotes the beginning of the continental slope. (3) Ocean basin outliers in which the basin-shelf separation is geologically and bathymetrically complex occur commonly in the Amerasia Basin and its separator from the Eurasia Basin, the Lomonosov Ridge. These are the furthest away from land and of uncertain geological character. Only because it is known that some of them are detached continental fragments that may contain suitable host sediments are they considered as a potential play.

Keywords Continental slope • Trough • Natural gas hydrate • GHSZ • Shelf depression • Basin outliers • Turbidite • NGH

We identify three different types or 'plays' of oceanic NGH in the Arctic that relate to marine sediments that have not been compacted since deposition in any other way than compression under the weight of younger sediments. NGH may also occur with older sediments within the continental shelves, but the geological framework of these will be different from relatively more modern sediments. The NGH petroleum system would be the same for each of the three plays, but their geography/geomorphology and location with respect to the shelf break are different. In particular, in relatively lower pressure shallow water, where temperature NGH differences of even a few °C have a strong impact upon NGH stability and GHSZ thickness (Fig. 3.1a). The very cold Arctic seafloor temperatures have the effect of bringing potential NGH high-grade deposits into shallower water depths than would be the case in open ocean marine sediments. Possible NGH in deep abyssal regions are not considered to have an immediate enough commercial significance to consider at this time.

(1) Continental margins. These are located in the uppermost part of the prism of continental slope marine sediments along continental margins. Seafloor water depths and distances from shore will be about the same as for deepwater conventional hydrocarbon deposits. In many cases these NGH concentrations may structurally overlie conventional hydrocarbon deposits. Sediments in continental margins generally have a continental provenance, often with considerably mature sorting of sediment types prior to deposition on the marginal sedimentary prisms. They may be deposited as deepwater turbidites, with overbank deposits, channel sands and basin fill, and other depositional types.

(2) Intra-shelf troughs/depressions/deep shelf margins. These sediments may have some of the characteristics of the shallowest deepwater turbidites found in the continental margins. They appear to be related to glacial streams and erosion by ice in the Barents-Kara (Zone 6.e) and Canadian Islands (Zone 6.1) outer shelves (Fig. 9.1) Sediments in them may be transitional between deeper water turbidites and shelf sediments and could be expected to have considerable coarse clastics deposited as the ice caps broke down and sea level rose. These sediments could have characteristics of continental shelf sediments that would be excellent hosts for NGH.

The GHSZ thicknesses shown in the troughs by Long et al. (2008), Wood and Jung (2008) data added an extra dimension and a new gas play to that envisaged by Max and Lowrie (1993), who did not have access to data of sediment deposited in the troughs in the glaciated continental shelves, but rather focused on the continental margins and deeper water marine sediments. Because of the extreme cold of the Arctic Ocean bottom waters, which are usually below zero year round, GHSZ reaches 150 m thickness over very large areas and up to 300 m over considerable areas, particularly in the Amundsen Trough on the Canadian margin and the St. Anna Trough on the Barents Sea margin (Long et al. 2008, Figs. 1 and 5).

These troughs may be prospective for NGH. In comparison to the Amundsen and Anna Troughs, where water depths are between 500 to over 1,000 m over a large area, the Nankai Trough has water depths of about 720–2,000 m and the base of the GHSZ indicated from BSRs seen on reflection seismic profiles varies from 177 to 345 m bsf (Birchwood et al. 2010). A discovery well at Nankai in 945 m water depth with a GHSZ thickness of 200–270 m showed the presence of considerable high-grade NGH (Takahashi et al. 2001). Large areas of intra-shelf troughs also occur elsewhere. For instance, the New Brunswick-Newfoundland-Labrador continental shelf that is washed by the cold Labrador Current has intra-shelf troughs and depressions potentially hosting GHSZ of up to 200 m thick equal in area to a substantial proportion of the area of continental slope sediments (Halliday 2011).

We note that the area of these plays appear to be a substantial fraction of the area of the continental margin zone likely to host GHSZ. Although the continental margin GHSZ will be thicker in deeper water than is found in these plays, the

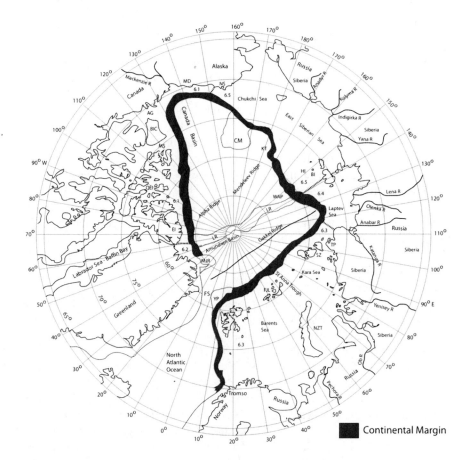

Fig. 9.1 Continental margin zone. Polar projection location map after (Jakobsson et al. 2004, 2008)

NGH accumulations in these intrashelf troughs may be thick enough to host substantial high-grade NGH. The potential for NGH concentrations in this play is currently unexamined, but we regard it as extremely promising. Because of the shallower water depths, exploration and drilling may be easier and less costly than in deeper water.

(3) Ocean basin outliers. These are areas of bathymetric highs oceanward from the conventional continental margin sedimentary prism drape from the shelf break. In the Arctic Ocean, the main outliers are the Yermack Plateau, the Chukchi microcontinent, the Morris Jessup Rise Region, and the Lomonosov Ridge. Because they are related to the opening of the Arctic Ocean basins they probably once have had sediment deposition of shelf and then continental margin character. As with the troughs, a study of their depositional environments will throw light on their capability to host concentrated NGH. These are not considered to be of first order

exploration interest, but could host considerable NGH. Their potential for NGH is currently unexamined, but possible.

The Continental Margin Zone (Fig. 9.1) is the most likely to contain NGH deposits of commercial value. This is the zone in which continental slope deepwater turbidite sediment similar to those hosting both conventional deepwater hydrocarbons and relatively complete NGH evaluations (Nankai and GoM locations) occur. Thus, proven technology can be applied directly. Our estimates for NGH (Table 10.1) have been made by extrapolating directly analogous, well-understood geological factors, and the special circumstances for the localization of NGH accumulations.

We have carried out a preliminary NGH petroleum system assessment of the likelihood of NGH resource potential in Arctic deepwater continental margin sediments, incorporating the regions of GHSZ likelihood defined by Max and Lowrie (1993) and the map of GHSZ thickness (Wood and Jung 2008) that show the general areas that we feel would be productive for exploration (Fig. 9.2). These are based on a first-order assessment (in which confidence limits are necessarily broad) that has integrated the likelihood of continental margin (including troughs and outlier) sediment source and host types, potential gas flux, turbidite depositional environments, GHSZ, and other factors. We have assumed a linear extrapolation of turbidite deposition across the continental margin sediment zone. Other continental margin sediments beyond the ~100 km limit and abyssal sedimentary basins may have potential for NGH and thick GHSZs.

In contrast to Max and Lowrie (1993), who identified a number of zones based on criteria for gas generation and sequestration within GHSZ, the zones shown in Fig. 9.1 refine that to identification of the likelihood of high-grade NGH host sediments. Although the broader development of NGH in the Arctic basin shown by Max and Lowrie (1993) may remain valid along with a number of other total NGH estimates for climate and biological assessment, only the more restricted areas based on identification of NGH in potentially economic zones.

We regard the Trough Zone (Fig. 9.2) as the next most likely repository of commercial NGH deposits. Troughs could host NGH both in unconsolidated sediments and in subjacent partly or fully lithified sediments. Although little is known about NGH potential in the Trough Zones, we regard the likelihood of NGH concentrations in them as good. A program to evaluate their NGH potential could prove very useful to NGH exploration as the water depths are generally shallower than the Continental Margin zone and relatively mechanically strong and highly permeable clastic sediments may dominate. The Outliers (Fig. 9.2) are those areas within and bordering the Amerasia and the Eurasia Basins (Fig. 1.1) and seaward of the bulk of the Continental Margin sediments, although overlap occurs. These are generally in deeper water, further from a nearby coastline, and have the least well known geology. Knowledge of their geological history and NGH potential must be regarded as lying further in the future.

O'Grady and Syvitsky (2002), who ascribe Plio-Pliocene sedimentation predominantly to ice streams, describe three continental margin types based on increasing sediment supply. These are sediment starved margins, inter-sediment fan areas, and deep-sea fans, although even sediment starved margins can have

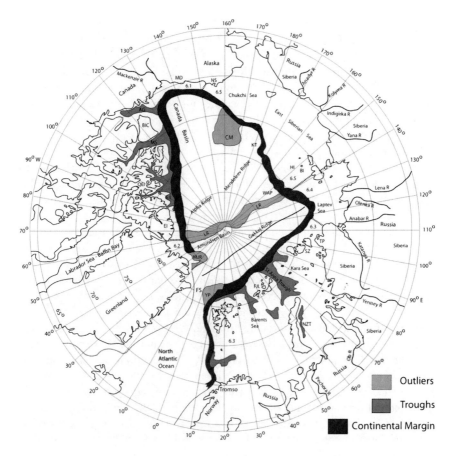

Fig. 9.2 The 3 principal oceanic NGH gas plays in the Arctic Ocean region. In the future, NGH concentrations in the abyssal and geologically more complex regions of the Eurasia and Amerasia Basins may have NGH potential. Polar projection location map after (Jakobsson et al. 2004, 2008)

turbidites and lower-slope sand lobes. Thicker sediments generally increase the likelihood of deepwater sands. O'Grady and Syvitsky (2002) also subdivided the circum-Arctic continental slope with respect to parameters likely to influence sediment delivery to the slope. They suggest that those continental margin regions that experienced higher sediment input generally dip more gently than slopes having less sediment input. They also found a direct relationship between the scale of troughs and the sediment fans on the continental margin. Longer troughs are associated with slope fans with a more gentle profile. This implies that the bulk of the sediment along Arctic continental margins is related to erosion and deposition during ice events. However, during interglacials, and along long stretches of continental margins of unglaciated land and shelf areas, fluvial and shallow water reworking of sediments are probably very important to forming sand bodies that find their way to deepwater.

Likelihood of NGH depends mainly on thickly sedimented zones that contain both source and host sediments, and GHSZ of suitable thickness. Generally, presence of thicker and more rapidly deposited sediment enhances the likelihood of gas generation by thermogenic or biogenic processes and the concentration of large volumes of NGH by processes within the sediment. Furthermore, rapid deposition of large sediment volumes also increases the probability of deepwater sands. From an economic point of view, the water depth is also important because drilling and production are more easily carried out in shallower water.

We have not established a water depth cutoff as an economic limit to potential NGH deposits because this will depend on commercial factors that can only be guessed at now. For instance, if NGH development has to await the spread of infrastructure established for conventional deepwater hydrocarbons, as is the case in the GoM, then the wait may be extensive. If development is to be made for NGH alone, as it will be apparently for the Japanese Nankai deposits, the economics of this in the Arctic are far from clear. It is also possible that a stranded gas solution will be applied. That is, where production of NGH would be commercially feasible but normal infrastructure transport to markets is not available, vessels with either gas to liquids, CNG, or NGH technology could increase the energy density of the transported gas enough to provide a commercial solution.

We follow Max and Lowrie (1993) in identifying almost the entire continental margin sediments of the Arctic Ocean deepwater (and the W. Barents margin) as having excellent potential for high-grade NGH concentrations that could be of economic character. It remains to be established whether margins that were subject to ice-related erosion and deposition and those dominated by fluvial process are significantly different in either their gas generating or NGH high-grade host development. In addition to the continental margins, we recognize the incised troughs of the glaciated marginal zones as also having good potential to host high-grade NGH concentrations. For instance, the St. Anna Trough and a number of other troughs in the Barents Sea are shown by Wood and Jung (2008) as having GHSZs up to 150 m and in some places up to 200 m in thickness, which is not dissimilar to the deep shelf/GHSZ of the Nankai deposits. The Canadian Arctic shelf is even more broadly incised with deep troughs that have GHSZ commonly in the 100–200 m range. The coalesced incised troughs in the Queen Elizabeth Islands have a broad, deep continental shelf margin, which considering the more extreme cold of the seawater, results in a thick GHSZ, not dissimilar to that at Nankai.

In addition to the continental margins whose slope break with the shelves defines their landward limit, there are a number of margin outliers that could also host NGH concentrations. The Yermak Plateau off Svalbard and zones in the Chukchi microplate were previously picked by Max and Lowrie (1993) as potential NGH areas. These areas are bathymetrically relatively shallow regions and thus have potential for exploration.

We explicitly rule out thickly sedimented regions in abyssal regions as being first order NGH exploration targets. Even though broad areas of thick GHSZ are shown in the abyssal regions of both the Amerasia and Eurasia Basins by Wood and Jung (2008), industrial technology will have to be improved

9 NGH Likelihood in the Arctic Ocean

Table 9.1 Probability potential of NGH in continental margin and intra-shelf troughs

Margin/zone	Continental margin	Intra-shelf depression	GHSZ thickness (m)	NGH likelihood
Alaska to Greenland	6.1		>400 m*	Excellent
Amundsen Trough		Trough	to 150 m	Excellent
M'Clure (Strait) Trough		Trough	150–250	Excellent
Q. E. Islands trough margins		Coalesced troughs	250–>300	Excellent
N. Greenland	6.2		300–400	Possible
Barents-Kara Seas	6.3		400	Excellent
West Barents margin	6.3.1		>300	Good
NW Barents margin	6.3.2		to 350	Good
St. Anna Trough		Trough	150–300	Excellent
Other Barents-Kara Intra-shelf troughs		6.3.3	100–150**	Excellent to good
Canadian intra-shelf troughs				
Kara Sea—E. margin	6.3.4		300–350	Good
Laptev-W. Siberian Seas	6.4		up to 350	Excellent
E. Siberian Sea	6.5		up to 350	Good
Outliers				
Yermack Plateau			to 200	Possible
Chukchi micro continent			300–350	Likely
Morris Jessup Rise region			200–300	Unknown
Lomonosov Ridge (flanks)			to 350	Unknown

Note All GHSZ thicknesses based on Wood and Jung (2008). Numbers in column 2 refer to text descriptions. GHSZ thicknesses generally applied from Wood and Jung (2008) and are used conservatively. Ocean basin potential NGH outliers such as the Yermak and Chukchi microcontinent are not shown but remain development possibilities in the future. Abyssal regions of thick GHSZ in sediments such as in the Amundsen Basin section of the Eurasia Basin are not shown, as they are not considered to have near-term economic NGH potential., The GHSZ thickness in the 6.1 zone should be thicker and more regular than shown by Wood and Jung (2008). This should be an area of very thick GHSZ owing to relatively low geothermal gradient and probably very thick sediment, thinning to the NW. This zone includes depressions that do not reach the continental margin, as well as troughs that do.

considerably before these regions could be considered as potential sources of natural gas. However, considerable NGH may be present, even if there is no likelihood of the presence of conventional petroleum systems. Table 9.1 summarizes our view of the likelihood of NGH concentrations in the Arctic Ocean deepwater.

References

Birchwood R, Dai J, Shelander D, Collett T, Cook A, Dallimore S, Fujii K, Imasato Y, Fukuhara M, Kusaka K, Murray D, Saeki T (2010) Developments in gas hydrates. Oilfield Rev 22(1):18–33

Halliday J (2011) Newfoundland and Labrador gas hydrates research program—A step in the right direction. Fire in the Ice 11(2) National Energy Technology Laboratory, U.S. Department of Energy, 14–17

Jakobsson M, Mcnab R, Cherkis N, Shenke H-W (2004) The international map of the Arctic Ocean (IBCAO). Polar Stereographic Projection, Scale 1:6,000,000. Research Publication RP-2. U.S. National Physical Data Center, Boulder, Colorado 90305

Jakobsson M, Macnab R, Mayer L, Anderson R, Edwards M, Hatzky J, Schenke H-W, Johnson P (2008) An improved bathymetric portrayal of the Arctic Ocean: implications for ocean modeling and geological, geophysical and oceanographic analyses. Geophys Res Lett 35(5):L07602. doi:10.1029/2008GL033520

Long PE, Wurstner SK, Sullivan EC, Schaef HT, Bradley DJ (2008) Preliminary geospatial analysis of Arctic Ocean hydrocarbon resources. U.S. Department of Energy/Pacific Northwest National Laboratory PNNL-17922

Max MD, Lowrie A (1993) Natural gas hydrates: Arctic and Nordic Sea potential. In: Vorren TO, Bergsager E, Dahl-Stamnes ØA, Holter E, Johansen B, Lie E, Lund TB (eds.) Arctic geology and petroleum potential, proceedings of the Norwegian petroleum society conference, 15–17 Aug 1990, Tromsø, Norway. Norwegian Petroleum Society (NPF), Special Publication 2 Elsevier, Amsterdam, 27–53.

O'Grady DB, Syvitski PM (2002) Large-scale morphology of Arctic continental slopes: the influence of sediment delivery on slope form. Geological Society London, Special Publications 203, 11–31. doi: 110.1144/G!G! SL.SP.2002.203.01.02

Takahashi H, Yonezawa T, Takedomi Y (2001) Exploration for natural hydrate in Nankai trough wells offshore Japan. In: Proceedings of the offshore technology conference 2001, Houston, USA, Paper. OTC 13040

Wood WT, Jung WY (2008) Modeling the extent of Earth's Marine methane hydrate cryosphere. Proceedings of the 6th international conference on gas hydrates (ICGH 2008), Vancouver, British Columbia, Canada, 6–10 July 2008

Chapter 10
Estimates of the NGH Resource Base in the Arctic Region

Abstract The natural gas potential of the Arctic Ocean NGH resource base is very large at 6,000+ TCF gas-in-place. This estimate has been made using a NGH petroleum system approach for the deepwater NGH resource of the continental slope sediments as part of a worldwide estimate of continental slope and deep continental shelf edge sediments. Estimates have been made for a zone extending about 100 km outward from the slope break. The calculation of gas-in-place for the continental shelf play in the Arctic Ocean region used a previously determined GHSZ thickness, estimates of suitable host sediments, water depths that are workable using today's deepwater technology, and information from three analogs (Nankai, Gulf of Mexico, various permafrost NGH deposits in Canada and Alaska). This estimate does not include the estimated abundance of NGH in lower grade deposits, particularly dispersed NGH in muddy sediments or other concentrations of NGH such as those known from vein-type deposits in fractured shales.

Keywords Continental slope • 100 km zone • Natural gas hydrate • GHSZ • Shelf depression • NGH analogs • Turbidite sands • NGH

The potential of the resource base of gas derived from NGH is very large. The detailed assessment conducted for the GoM, North Slope of Alaska, and Nankai areas have been used as a basis for the broad estimates of the NGH resource base (Johnson 2011, 2012). Using a NGH petroleum system approach, this has allowed the estimates of the deepwater NGH resource to be based on statistical likelihood of suitable high-grade NGH in sand hosts as the primary economic targets. No estimate has been made for fracture-hosted deposits, as they are not considered to be a first-order economic resource potential. The Nankai NGH deposit to the SE of Tokyo also occurs in sand bodies of turbidite channel origin (Noguchi et al. 2011). Their concentration ranges appear to be similar to those of the deepwater Terrebonne Basin in the central GoM (Frye 2008).

The areal NGH resource density within the turbidite-related sands of the Terrebonne Basin is calculated at 1.183×10^9 m^3 per km^2 where delineated sand reservoirs are present and 0.32×10^9 m^3 per km^2 where sands are thought to be absent (Frye et al. 2011). These results have been used to calculate the NGH resource in high-grade sands, using estimates for sand %. With essentially no data for sediment composition through an anticipated GHSZ in the Arctic Ocean Basin, but with comparisons with some other continental margins available, sand percentages in the GHSZ probably vary between 5 and 10 %, with excursions to both lesser and greater percentages from place to place depending on the depositional environment.

The estimate for the NGH resource base in the Arctic is based on a number of assumptions similar to those commonly used in the energy industry (Johnson 2012). Estimates are only for gas in NGH, with no estimates made for trapped gas related to the NGH. These estimations were carried out initially for the Global Energy Assessment (GEA) coordinated through the International Institute for Applied Systems Analysis (IIASA) based in Vienna, Austria. Further calculations for economically and theoretically recoverable gas were made.

1. NGH region: Estimates were only made for a zone about 100 km basinward from a shelf break (continental slope) or from a deep shelf area greater than about 800 m water depth (depressed shelf margin or trough).
2. The GHSZ thickness within this region were determined from seismic interpretations or estimated based on structural position and likely heat flow.
3. Host sediments: Conservative estimates were made for the proportion of turbidite sands within the NGH region based on known examples and extrapolation using industry-standard methods. We have not applied a numerical approach such as described by Felletti and Bersezio (2010), but have based our estimates on turbidite system characteristics of a high sediment continental margin similar to the GoM. While sands are transported further basinward with turbidity currents (as documented by the various drilling programs), using a particular breakout width for the likelihood of a continental margin NGH zone allows for a better averaging of sand content for the sediment volume.
4. Workable water depths: The arbitrary 100 km wide zone may be narrower. Water depths greater than those in which drilling and production are carried out, or are liable to be carried out within about a 10 year window, are encountered. This keeps the results in water depths where operations are (currently) feasible.
5. Enough is now known from three locations, Nankai. GoM, and drilled permafrost NGH deposits to provide ranges of NGH saturation consistent with seismic interpretation of NGH in-place and concentration.
6. Gas-in-Place calculated: After multiplying the various ranges in value for sand %, within the GHSZ (including a range in values for the GHSZ volume derived from the work of Wood and Jung (2008), NGH saturation estimates using an average cell saturation approach, gas-in-place figures were derived (Table 10.1).

Table 10.1 Estimates of NGH gas-in-place in sands (High Grade NGH deposits) for Arctic Ocean continental margin zone only by country and region. After Johnson (2012)

Country/region	Range of estimates (TCF)	Median (TCF)
Canada	533–8,979	2,228
Western Europe (including Greenland)	36–14,858	1,425
U.S. Arctic *onshore* technically recoverable (USGS 2008)	No separate offshore estimates	85.4 TCF
FSU (Russia)	1,524–10,235	3,829
Arctic Ocean	178–55,524	6,621
World	4,705–313,992	43,311

The upper values of the ranges of estimated gas volumes describe an upside that would be achievable with new technology that could be developed in the next 30 or 40 years. For technically recoverable gas, 50 % of "in place" median was used, with a range of 25–75 %. The economically recoverable gas volume was assumed to be an order of magnitude less than the technically recoverable volume, with zero being used as a low end. Without a sustained commercial test it is possible that none of the NGH is currently economically viable (depending on who is doing the economics). To reduce economics to a common base, the assumption was made that no government subsidies or special treatment for any resource would be available.

It is likely that considerable NGH also occurs in continental shelf troughs and possibly on the ocean basin outliers, but no estimates have yet been made for these areas because the technology required is not yet available. Even though there is a considerable area of suitable GHSZ, especially in the continental shelf outer troughs, these are a different depositional environment from the well-understood continental margin turbidites systems. These may require a study in their own right.

These estimates of gas-in-place are very large when compared with natural gas usage. 23 TCF is the annual gas consumption by the US, with a 2010 world gas consumption of 113 TCF. Most important, the Japanese 3.7 TCF consumption (and their proposal to have 1 TCF per year from NGH when Nankai is fully developed, with production scheduled to begin in 2013) may be further augmented by developing other indigenous Japanese NGH resources in order to achieve the universally commendable aim of energy independence.

As much as a quarter of the world's remaining undiscovered conventional oil and gas might reside in the Arctic region, more than 80 percent of it in deep water (Collett et al. 2008). The suggestion by Max and Lowrie (1993) that the Arctic basin contains a very promising NGH resource because of its thick continental flanking sediment, their gas generating potential, and widespread thick GHSZs, seems increasingly well founded. The amount of natural gas within the NGH accumulations of the world is believed to greatly exceed the volume of known conventional natural gas reserves (Fig. 6.1). It is not known how much or what percentage could be technically or economically recoverable.

The history of the hydrocarbon industry, however has been to improve production from original deposit estimates, and these improvements can be expected in unconventional hydrocarbons also. Although oceanic NGH is considered to hold 95 % of the world's NGH (Kvenvolden and Lorenson 2001) much more is known about onshore subsurface permafrost NGH because many of these deposits have been drilled and evaluated along with their related conventional gas and oil deposits. Estimates for median Arctic NGH resources (Johnson 2012) are almost 15 % of the world NGH estimate (Table 10.1). Overall NGH resources may dwarf those of conventional hydrocarbon deposits and the Arctic may be the single richest NGH region on Earth as a function of overall area, but NGH in sands, which is the primary high grade host for exploration, is about equivalent to the sum of projected conventional gas resources (Fig. 6.1).

The theoretical NGH gas resource may prove to be considerably larger in the future. If new drilling, conversion, and exploration technology would make smaller high grade and medium grade deposits, fracture fill, and especially the low grade deposits that constitute by far the most abundant component of NGH. Some of these, beginning with deposits having some characteristics of the high grade deposits in the 'hydrate pyramid' (Fig. 6.1) and then proceeding 'downward' in the pyramid as the low grade deposits become increasingly less like the high grade deposits in terms of % NGH and geotechnical characteristics. In some geological situations this level of natural gas sequestration may be more important for consideration in global warming scenarios than for economic considerations. Therefore, research into the broad range of NGH occurrences and concentrations may have important application in fields other than economic NGH production.

References

Collett TS, Agena WF, Lee MW, Zyrianova MV, Bird KJ, Charpentier RR, Cook T, Houseknecht DW, Klett RR, Pollastro RM, Schenk CJ (2008) Gas hydrate resource assessment: North Slope, Alaska. USGS fact sheet from October 2008. http://geology.com/usgs/alaska-gas-hydrates.shtml. p 3

Felletti F, Bersezio R (2010) Validation of Hurst statistics: a predictive tool to discriminate turbiditic sub-environments in a confined basin. Pet Geosci 16:401–412. doi:10.1144/1354-079309-005

Frye M (2008) Preliminary evaluation of in-place gas hydrate resources: Gulf of Mexico Outer Continental Shelf. U.S. Department of the Interior Minerals Management Service Resource Evaluation Division OCS Report MMS 2008-0004, p 136

Frye M, Shedd W, Boswell R (2011) Gas hydrate resource potential in the Terrebonne basin, Northern Gulf of Mexico. Marine and petroleum geology, p 19. doi:10.1016/j.marpetgeo.2011.08.001

Johnson AH (2011) Global resource potential of gas hydrate—a new calculation. Fire in the Ice. NETL, U.S. Department of Energy 11(2), 1–4

Johnson AH (2012) Gas hydrate. In: GEA, 2011: the global energy assessment. IIASA, Laxenburg, Austria and Cambridge University Press, Cambridge. pp 35–43

Kvenvolden KA, Lorenson, TD (2001) The global occurrence of natural gas hydrate, in natural gas hydrates: occurrence, distribution, and dynamics, In: Paull CK, Dillon WP (eds) American Geophysical Union Monograph vol 124, pp 3–18

Max MD, Lowrie A (1993) Natural gas hydrates: Arctic and Nordic Sea potential. In: Vorren TO, Bergsager E, Dahl-Stamnes ØA, Holter E, Johansen B, Lie E, Lund TB (eds) Arctic geology and petroleum potential, proceedings of the Norwegian petroleum society conference, 15–17 Aug 1990, Tromsø, Norway. Norwegian Petroleum Society (NPF), Special Publication 2 Elsevier, Amsterdam, pp 27–53

Noguchi S, Furukawa T, Aung TT, Oikawa N (2011) Reservoir architecture of methane hydrate bearing turbidite channels in the eastern Nankai Trough, Japan. Proceedings of the 7th international conference on gas hydrates (ICGH 2011), Edinburgh, 17–21 July 2011, p 9

USGS (2008) Gas hydrate resource assessment: North Slope, Alaska, USGS Fact Sheet, p 2. http://geology.com/usgs/alaska-gas-hydrates.shtml

Wood WT, Jung WY (2008) Modeling the extent of Earth's marine methane hydrate cryosphere. Proceedings of the 6th international conference on gas hydrates (ICGH 2008), Vancouver, British Columbia, Canada, 6–10 July 2008

Chapter 11
Oceanic NGH: Low Risk Resource in Fragile Arctic Environment

Abstract Risk to the environment traditionally consists of an uncontrolled leakage of a gas or oil, resulting in pollution of the environment. Exploration, including drilling, and production of NGH carries an extremely low risk worldwide. The very low environmental risk is particularly important in the environmentally fragile Arctic environment. We suggest that the very low environmental risk factor may be key to the development of NGH in the Arctic Ocean region. When NGH is converted for production, only relatively pure methane and water are produced. NGH is stable within its reservoir and will not convert to its constituent gas and water unless the formation pressure is lowered or the temperature is raised sufficiently to introduce instability conditions. Thus, with careful drilling, the danger of gas venting is very low. In addition, during production the maximum gas pressures in the reservoir can be controlled, in strong contrast to conventional gas deposits in which very high pressures may exist from the outset. In addition, NGH deposits are not associated with liquid petroleum (oil), especially in the Arctic where predominantly biogenic NGH can be anticipated. Thus, even if a gas leak occurs, virtually no environmental hazard to macrofauna such as birds and Arctic mammals exists.

Keywords Low risk • Environment • Natural gas hydrate • GHSZ • Pure gas • Low pressure • Oil • NGH

In winter the floating sea ice covers almost all of the Arctic Ocean, and extends into the North Atlantic Ocean and Labrador Sea. But the Arctic region is presently undergoing the most dramatic warming of a warming world. The areal extent of the summer sea ice, which recently has shrunken less than 85 % coverage of the entire ocean, tends to pack along the Canadian and Greenland Arctic Ocean margin and usually leaves considerable expanses of ice-free water along the Eurasia and Alaska—western Canada margin.

The Arctic Ocean is an environmentally and ecologically fragile region because of the low temperatures and the dispersed ecosystem that has adapted to the environment (Larsen et al. 2001). Because thin single year ice is becoming more common at the expense of thicker multiyear sea ice and the percentage of ice cover is lower earlier in the spring and later in the autumn, ship traffic is increasing. Most important, however, hydrocarbon exploration is now moving into the Arctic because warming is bringing less inhospitable working conditions and a longer working season.

The high Arctic is the current frontier for hydrocarbon exploration, and will remain so for some time (ACS 2013). In the future, the South Polar Region may also attract hydrocarbon exploration, but first the resources of the high Arctic, which are much closer to major markets, must be mastered. Large oil and gas discoveries on land in Russia and Alaska date from the 1960s. More than 35 fields are now in production across Alaska, Russia, Norway (Barents Sea) and Canada. Regulatory agencies and courts have slowed exploration because of environmental concerns. Now, with the imminent likelihood that the ocean ice cover will significantly diminish or even disappear in the summers within decades, the hydrocarbon resources of the deeper water Arctic areas are beginning to be assessed. Shell, for instance, has invested about $4 billion on 10-year leases and has been trying for at least 5 years to drill in Alaska's Chukchi and Beaufort Seas. But regulatory agencies or courts have delayed exploration efforts, because of concerns that Arctic waters are vital breeding grounds for many aquatic species that are endangered or at risk and that a well blowout could cause a huge release that would be difficult or impossible to control.

Not only is the environment difficult for exploration, but because of the cold, the operation of natural 'cleanup' chemical and biological systems that can act to remediate petroleum spills work very slowly and industrial cleanup can be significantly hampered by the difficult logistics, working conditions, and freeze up of equipment that endeavors to separate petroleum from water in sub-freezing conditions. Because of the general lack of natural remediation, the Polar Region is generally regarded as being the most environmentally fragile on Earth. The Arctic Ocean, which is centered on an ocean basin surrounded by sparsely inhabited land, is the subject of immediate environmental concern. Even relatively small spills can lead to immense damage to the environment and its ecology in the environmentally fragile Arctic region.

Because the Arctic is essentially an enclosed sea surrounded by sophisticated nations with respect for rule of law, it is envisaged that issues arising from exploration for and production of the hydrocarbon resources of the Arctic will be dealt with rationally by the Arctic and other nations. The recent agreement between Norway and the Russian Federation as to their median line across the Barents-Kara Sea shelf and the agreement of the Arctic Nations to agree to binding arbitration under UNCLOS III (Connerty 2006) for establishing natural resource boundaries in the Arctic Ocean are encouraging signs of the peaceful solution to national energy issues in the Arctic region. The most important organization of states in the Arctic region is the Arctic council, which consists of those states bordering the Arctic and observers with a different status, under which most international agreements for the Arctic are promulgated (Arctic Council 2013).

11.1 Risk Factors of Conventional Hydrocarbon Production

Pressurized gas and liquid petroleum fills pore space in conventional reservoirs trapped by an impermeable seal and held in place by buoyancy over subjacent pore water. Because of its pressurization, the hydrocarbon material contains immense potential energy; a veritable genie in a bottle that must be released slowly and with great care during production. Very large amounts of gas and petroleum may be concentrated in hydrocarbon deposits that may have been stable for long periods of geological time. Generally, gas and oil may occur together in the same reservoir or in the same system of traps related to migration pathways from subjacent sources. As conventional reservoirs often reside at considerable depths beneath the seafloor and are at or above the ambient pressure for that depth, the pressure differential with the seafloor and the surface is usually very large.

The natural pressurization of conventional hydrocarbon deposits serves a very practical purpose, that of inexpensively driving the hydrocarbons to the surface under controlled production conditions. This natural flow from the reservoir to the surface is referred to as 'primary recovery'. When the natural drive diminishes, secondary recovery techniques involving chemical, thermal, solvent, pumping, fracking, or other stimulation may be required to increase pressure, reduce viscosity, open porosity and increase the flow of hydrocarbons from the reservoir to the surface. All secondary recovery stimulation techniques have additional cost factors.

The naturally dynamic conventional hydrocarbon deposits, however, present a continuing problem, particularly during exploration and early production. Because of their high pressures, conventional gas and petroleum deposits have the potential to uncontrollably vent if a breach occurs in an exploratory or production drilling system, pipeline, or other infrastructure. And because of their high temperatures, they can be difficult and dangerous to handle. Regulation, development of best practices, new deepwater equipment, and simply the high costs of dealing with blowouts may strongly reduce the risk of blowout, but a risk remains nonetheless. If a conventional deepwater hydrocarbon would be exposed to the overlying seafloor water, it would likely spontaneously blow out.

11.2 Inherent GeoSafety of NGH Production

In contrast to conventional hydrocarbon reservoirs, NGH is stable and effectively inert in its reservoir. It exists at the ambient pressures and temperatures at any depth in the GHSZ. NGH is a thermodynamically stable solid in its natural environment and is unlikely to be spatially associated with petroleum. Gas cannot be recovered from NGH without artificial stimulation that alters ambient conditions within the GHSZ. If NGH deeper in the GHSZ were to be exposed to overlying seafloor water, it would only become more stable as a solid crystalline material because it would be further cooled. NGH can thus be regarded as an environmentally secure resource. In order for gas to be produced, the NGH must be artificially stimulated so that it may be converted to its component gas and water or dissolved as the first step of a gas production methodology. NGH will not naturally dissociate to its component gas and water so long as stability conditions are maintained, even in drill bores. Extraction of NG from NGH must begin with one or more of a number of processes that would be considered as a secondary recovery technique in conventional hydrocarbons.

There are four main methods for NGH conversion, all of which can be accomplished using existing or emerging technology. These include thermal stimulation, depressurization, dissolution, and chemical exchange (Max and Johnson 2011). Each approach has advantages and disadvantages related to operating expense, attainable flow rates, and volumes of produced water. Determination of the optimal approach will depend on specific reservoir conditions, costs, potential environmental impact, and other considerations. The first production test of NGH production in the Nankai deposit in March, 2013 used a specially designed electric submersible pump system able to depressurize the test section and to separate natural gas from water and move them to the drillship through separate production strings (OGJ 2013). Baker Hughes Inc. designed the production system for the Japanese research consortium.

NGH formation and dissociation is a chemical reaction that produces heat upon formation and consumes heat during dissociation. This introduces a natural buffering that acts to slow reaction rates. For instance, when NGH begins to form spontaneously, heat is produced that drives the reaction point in P–T space back toward the phase boundary. When NGH begins to dissociate, heat is consumed that tends to drive the reaction point back toward the phase boundary. Thus, NGH is naturally "self-preserving". NGH only disassociates at crystal boundaries with pore water where diffusional processes can actively transfer gas molecules from NGH to the pore water. The mass to surface area ratio is very important to controlled dissociation. In very concentrated, high-grade NGH deposits (Max et al. 2006) where there may be little remaining permeability, the creation of additional surfaces along which dissociation may take place may be necessary. As dissociation proceeds, permeability should increase. NGH does not have the potential to explosively decompress to its component gas and water, even if suddenly removed to pressure–temperature reaction points in which the NGH is very unstable, a point made obvious by the many images of NGH cores being examined on the decks of drilling ships.

The most important safety factor in any recovery scenario from the marine NGH system is that not only is solid crystalline NGH physically stable within the GHSZ at the ambient pressure at which it occurs, but if either the pressure or temperature conditions are changed to those of instability, the natural buffering of the reaction system tends to slow dissociation reactions. Because it is stable, even if a natural or manmade pathway to the seafloor or the surface is made, no gas will evolve from the NGH so long as it remains at ambient pressures and temperatures. Unplanned stimulation of NGH, for instance by adding heat or inhibitors as part of a drilling process to a section containing unrecognized NGH, could cause unwanted gas to evolve.

11.2 Inherent GeoSafety of NGH Production

We would suggest, for instance, that an unplanned gas surge during drilling that could have a NGH source should trigger immediate remedial action such as stopping any drilling practice that might cause dissociation of NGH, and including injection of cold fluids to restore the ambient conditions. Because NGH conversion can only take place following the imposition of NGH instability conditions, removal of these artificial conditions, combined with the natural buffering of the reaction system, will rapidly slow conversion and possibly even reintroduce conditions of NGH stability in which gas molecules will again begin to be incorporated in NGH in the reservoir.

NGH occupies host sediment pore space as a solid material and, as a result, porosity and permeability will increase as a function of NGH dissociation. As gas is removed from the reservoir during gas production, pressures in the reservoir can be controlled by balancing NGH conversion with gas removal. Gas pressures can be both increased and decreased at will. Gas pressures should only be high enough to achieve commercial extraction rates (Fang and Lo 1996) in order to keep the pressure and the amount of free gas in the reservoir within safe limits. Water will flow into only some of the space occupied by the converted NGH because each m^3 of NGH contains about 0.8 m^3 of water. Water cut may be lower in produced gas from NGH than from a conventional deepwater gas deposit, especially if the reservoir itself can be used to initiate gas–water gravity separation.

In a highly pressurized conventional gas deposit, substantial gas is dissolved in the water and is available to form gas phase throughout a water mass upon production as pressure is lowered near the wellbore. This intermingling of free gas and gas evolving from water (and possibly oil) provides the drive but brings substantial water to the surface along with the gas. Ideally, because NGH deposits are close to the seafloor in geomechanically weak sediments either natural or artificial injection of water into the NGH production zone would be necessary to prevent the sediment host from compaction that could lead to sediment failure.

In the early stages of a NGH conversion operation, fracking or some analog process may be required to open porosity in very high-grade NGH deposits. Increasing NGH surface area would accelerate the rate of NGH conversion, especially early in the production process before solution pathways open along NGH grain boundaries. Higher pressures that cause faulting are of little danger to the environment in a NGH concentration because the higher pressures of the fracking will drive the NGH further into their stability field. Faulting associated with the fracking will not release gas because gas will not yet have been produced from converted NGH. If fracking is required after gas extraction has commenced, cessation of conversion stimulation will allow some, if not all, of the gas to reform NGH, stabilizing it. Without compensating for the heat requirement of a greater conversion rate during conversion and extraction operations, the two or three phase assembly of water, NGH, and gas will cool toward conditions of NGH stability, and gas conversion will slow or could even reverse, with the reformation of NGH.

Conventional deepwater hydrocarbon deposits have a maximum volume of naturally dynamic material during exploration and the early phases of production, which is when the risk of uncontrolled leakage is greatest. In contrast, the gas in

NGH is tied up in a very stable solid form. Gas must first be produced by conversion before it can be recovered or released. This renders natural NGH inherently resistant to uncontrolled venting during gas extraction operations. In the event of an incipient leak in the reservoir-production system, the deposit can be brought back to stability by cessation of the artificial stimulation. Early production of gas from NGH in the Arctic would pose an extremely low risk to the environment. New technology optimized for NGH exploration and production offers promise of being less expensive than for conventional petroleum and new NGH-specific regulations could spur near-term development.

Where a gas phase is generated in the reservoir through NGH conversion, the volume of gas in the reservoir can be kept small and its pressure controlled by matching conversion to extraction. With a low-gas-in-reservoir conversion-production methodology, the chance of blowouts from NGH production is dramatically reduced. Without large volumes of overpressured gas, significant blowouts cannot take place, even if there is a serious engineering failure. NGH conversion to gas is controllable, and must be controlled for production because of the relatively thin, semi-consolidated overburden. If a breach should take place, however, induced conversion can be stopped rapidly and the naturally buffered NGH conversion will cease, even if no remedial measures are taken. Thus, as a practical matter, NGH production may be considered to be essentially fail-safe.

Oceanic NGH tends to be relatively pure because it is dominated by biogenic methane produced at temperatures too low to allow higher density hydrocarbon gases or complex hydrocarbons (i.e., oil) to form. The natural gas produced from NGH appears to have about the same purity worldwide. In addition to the basic NGH-forming component, any chemical or dissolved ionic material migrating with the dissolved gas that is not a NGH-former is rejected from the crystallizing NGH into the pore water where it will dissolve and be carried away. These impurities appear to equilibrate and be removed. This is particularly true for salt rejected from the pore water. Low salinity zones were one of the primary indicators for the presence of dispersed NGH in the Blake Ridge (Paull et al. 1996). If the rejected salt had not been dispersed following formation of the NGH, there would have been no low salinity zones marking the presence of dissociated NGH. Liquid petroleum is also rejected by growing NGH, which may take up dissolved gas directly from the liquid petroleum (personal laboratory observation).

Biogenic NGH is not normally associated with liquid petroleum or gas condensates that are derived from thermogenic sources. In addition, NGH usually has little nitrogen, SOx, CO_2, and other contaminants that are often found in conventional deposits, and almost pure water is produced when NGH dissociates. Even if some converted gas and water were to leak from a NGH deposit, there would be no pollution, in the common sense of the word. Venting or leaks of gas from NGH conversion would have virtually no biological impact (except possibly to stimulate the base of the food chain) or visual effect.

Thus, there is a marked contrast between conventional hydrocarbon and NGH deposits, which has implications for the environmental risk factor of both exploration and production activities. Early production of gas from NGH in the Arctic poses an infinitesimally small risk to the environment. Accelerated development of the NGH resource can take place in the near term in an environmentally safe manner. Arctic deepwater gas exploration and production can begin safely with NGH.

References

ACS (2013) Kiruna declaration on the occasion of the eighth ministerial meeting of the Arctic council, Kiruna, Sweden, 15 May 2013, p 7

Arctic Council (2013) Senior Arctic officials report to Ministers. Arctic Council Secretariat, Kiruna, Sweden, 15 May 2013, p 144

Connerty A (2006) The international tribunal for the law of the sea and dispute settlement under UNCLOS III. file:///Users/michaelmax/Desktop/The%20International%20Tribunal%20for%20the%20Law%20of%20the%20Sea%20and%20Dispute%20Settlement%20under%20UNCLOS%20III%20-%20Part%20III.html

Fang WY, Lo KK (1996) A generalized well-management scheme for reservoir simulation. Paper SPE 29124 presented at the 13th SPE symposium on reservoir simulation, San Antonio, Texas, p 8

Larsen T, Nagoda D, Andersen JR (eds) (2001) The Barents sea Ecoregion. A biodiversity assessment. World Wild Life Fund, p 80

Max MD, Johnson AH (2011) Methane Hydrate/Clathrate Conversion. In: Khan MR (ed) Clean hydrocarbon fuel conversion technology. Woodhead Publishing Series in Energy no. 19. ISBN 1 84569 727 8, ISBN-13: 978 1 84569 727 3. Woodhead Publishing Ltd. Cambridge, pp 413–434

Max MD, Johnson A, Dillon WP (2006) Economic geology of natural gas hydrate. Springer, Berlin, p 341

OGJ (2013) Methane hydrate test used special ESP. Oil Gas J. http://www.ogj.com/articles/2013/05/methane-hydrate-test-used-special-esp.html?cmpid=EnlLNGMay282013 (23 May 2013)

Paull CK, Matsumoto R, Wallace P et al (1996) Proceedings of the ocean drilling program. Initial reports 164, Ocean Drilling Program, College Station, TX, p 623

Chapter 12
Economic and Political Factors Bearing on NGH Commercialization

Abstract Primary factors that will inhibit the production of natural gas from NGH are the 'stranded' character of any deepwater gas or oil from the continental slope regions of the Arctic Ocean and the absence of any production and transport infrastructure through which the gas can be brought to market. We suggest that special regulations governing NGH can be written that will allow for a lower level of risk than for any conventional gas resource while dispensing with much of the expensive regulation associated with conventional gas exploration and production. This would have an enabling effect on production from NGH through dramatically lowering the cost of exploration and production.

Keywords Stranded • Infrastructure • Natural gas hydrate • GHSZ • Regulations • Production • Transport • NGH

The primary general factor bearing on commercialization of NGH is whether produced gas can be sold at a profit level adequate to make it competitive with other gas resources. This competitive criteria applies to gas supplies generally and is not specific to any gas resource, be it conventional or unconventional. There are considerable price differences between regions and there is no international gas price comparable to the international oil price structure. Because of changing demand, gas prices shift and this creates an impetus for development of new resources, of which NGH is a prominent possibility. For example, the spot price for LNG went above $15/MMBTU in the autumn of 2011, making the Japanese program look far more viable. Two years ago there was debate on whether a $10 price in East Asia was sustainable. Part of the price jump is from Japan increasing imports by 14 % to cover power generation after the March 2011 earthquake and tsunami, and it is likely that this demand will increase as the politically disfavored nuclear power generation plants are replaced by gas.

Up to June 2012, the gas price in the UK has remained between $5.40 and $6.00 MMBTU, and as a result, many LNG tankers are rerouting from Europe to Asia. This may be a significant issue in early 2012 in Europe as new contracts for LNG are struck. Because of the development of shale gas and its promise for many decades of natural gas supply from indigenous sources, the US NGH program is probably all but dead, even though the Joint Industry Project (JIP) has gained results in the northern GoM funded largely by the U.S Department of Energy. With the US price now less than $5/Mft^3, there is little impetus for the U.S government to support non-academic research on NGH unless there would be a major disaster with shale gas production (Mooney 2011) that would result in severely constricting production. However, if the shale gas resource proves to be less than originally estimated (PennEnergy 2012), interest in NGH could restart development activity.

All produced conventional gas is not the same price and conventional gas is not necessarily less expensive to produce than some unconventional gas resources. For example, the giant Shtokman gas field (Zhdannikov and Mosolova 2007) that was on track to be brought into production as a joint Russian—Statoil (Norway) operation, with much of the gas scheduled for export as LNG, has been put on indefinite hold because of the abundance of gas available in the world (Andersen 2012). The Shtokman Field has always been a borderline cost competitive issue because of its high capital and operating costs and flow assurance issues in the 125 mile subsea pipeline. These cost factors compete with the lower cost conventional gas resources of the Yamal Peninsula on the Russian mainland immediately to the south coming on stream.

The world gas demand has been alleviated particularly in the United States, which is rapidly bringing its shale gas resource into production (Casselman 2009) and driving prices down to below $3/MMBTU, although the recent downgrading of the shale gas-in-place resource by the U.S Department of Energy in January 2012 may introduce further instability into the gas market. Shale gas production may soon be ramping up in Europe. Significant downgrading of shale gas resources from early exuberant predictions of the recoverable gas-in-place may allow renewal of interest in NGH.

If the Shtokman Field is not presently viable at current prices, NGH production is even further in the future for the Arctic, on a purely cost basis. There are offshore gas discoveries in the Kara Sea, not far from land, that are not yet being developed. However, if there is a solid test of NGH production in Japan, which is scheduled for the summer of 2012, and verification of producible NGH-bearing sands in India interest in Arctic NGH among the major energy companies and large independents may be renewed. It must be pointed out that a pure cost factor is rarely pure. For instance, exploration and production may be encouraged and even subsidized by nations on other grounds. Energy security is very important to all countries, especially to those with limited energy supplies, which are at the mercy of the international market, and natural or man-created disruptions in the energy supply trains. Energy security is the principle factor behind both the Japanese and Indian NGH programs, with early production desired well before gas available on the international market becomes scarce.

As well as the value or sale price of gas produced from NGH being unknown in the out years in which it might be developed, capital costs (CAPEX) and operating costs (OPEX) can only be estimated in relation to costs for existing conventional developments in the marine area. Although NGH exploration production factors, in particular seismic surveying and drilling, may allow lighter and much less costly technology to be applied (Sect. 8.3.3), other production issues will have exactly the same costs as for conventional gas. Flow assurance, for instance, will continue as a major concern. Transporting the gas, once it reaches the seafloor, will also entail the same costs because, "it's just gas". Once the gas has been converted from its solid NGH form and concentrated in a productive system, it has inherently the same issues as conventional gas except that NGH will almost always be pure natural gas with no associated natural gas liquids or oil.

In the Arctic, the continental margin zone hosting turbidite sands (Fig. 9.2) that are envisaged as the primary economic NGH play, are very far from shore and even further from markets, especially on the North American continent. Sediments in the outer continental shelf troughs (Fig. 8), which may have considerable affinity with deep outer shelf NGH deposits such as the Nankai Field SE of Tokyo, Japan, are slightly closer to shore and are in less ice-covered regions of the Arctic than along the North American margin. On a pure cost basis, it is unlikely that Arctic oceanic NGH will be developed any time soon. However, just as political imperatives have thrust Japan and India into the forefront of NHG development, it could be that the aspect of low environmental risk, which may inhibit conventional gas and oil development and lower costs of conforming to NGH regulations, which could be much less rigorous than for conventional gas and oil, may allow telescoping of NGH development timeline.

Alternatively, if the Japanese and/or Indian NGH programs begin continuous production of natural gas from their NGH resources, it is entirely possible that major companies will review their decision to put NGH development on hold. Major companies may want to reduce their risk by not wanting to be first in a new field, but they all will know a good resource play when they see one. Once commercial NGH production begins anywhere, it is likely that few major companies will want to be left out of a new and potentially very large resource play.

We suggest that special regulations governing NGH can be written that will allow for a lower level of risk than for any conventional gas resource while dispensing with much of the expensive regulation associated with conventional gas exploration and production. This would have an enabling effect on production from NGH through dramatically lowering the cost of exploration and production.

References

Andersen ES (2012) Personal communication to audience at meeting. Arctic Frontiers, Tromso, Norway, 25 Jan 2012
Casselman B (2009) U.S. gas fields go from bust to boom. Wall Street J (WSJ.com):5. http://online.wsj.com/article/SB124104549891270585.html#printMode
Mooney C (2011) The truth about fracking. Sci Am:80–85
PennEnergy (2012) U.S. shale gas reserve estimates plummet, p 1. http://www.pennenergy.com/pennenergy-2/en-us/index/articles/pe-article-tools-template.articles.pennenergy.petroleum.exploration.2012.January.u_s_-shale_gas_reserve.html
Zhdannikov D, Mosolova T (2007) Russia's Gazprom ups Shtokman reserves to 3.8 cm. Reuters. http://www.reuters.com/article/2007/11/15/gazprom-shtokman-idUSL1589543420071115

Chapter 13
Logistical Factors for Arctic NGH Commercialization

Abstract Liquid petroleum oils can be efficiently transported because their energy density is very high and they remain liquid even well below freezing. Gas is less economic to transport on a BTU basis because it has a lower energy density in any of its transport forms. Compressed natural gas (CNG), liquefied natural gas (LNG), and artificially manufactured gas hydrate (MGH) all have energy densities much lower than oils. The only comparable energy density for natural gas is artificially produced petroleum.

Keywords CNG • LNG • Pipeline • Manufactured gas hydrate • Stranded • Gas-to-liquid • GTL • NGH

Because of the remoteness of any NGH concentration in any of the three High Arctic NGH play areas that may be large enough to consider for commercial production (Fig. 9.1), the lack of existing pipelines and other infrastructure introduces a strongly negative factor to development economics. NGH in the Arctic will thus have the character of 'stranded gas', which is an industry designation given to gas deposits in which the cost of transporting the gas to market is a strongly prohibitive factor. The long pipelines and other fixed infrastructure necessary to bring High Arctic gas and oil to shore and to market will probably be too expensive for some time using current cost models. All High Arctic hydrocarbons would currently be regarded as stranded; this is not an issue for NGH alone. All other hydrocarbon resources in the High Arctic currently suffer from this same issue.

In locations such as the GoM, NGH development can ride on the coattails of existing conventional hydrocarbon exploration, equipment development and deepwater technology and techniques. In the Arctic Ocean, conventional fixed hydrocarbon infrastructure may not reach any of the three NGH prospective zones (Fig. 9.2) until after continental shelf hydrocarbon development progresses slowly poleward into deeper water. Establishment of infrastructure will follow exploration and production opportunities. And this may not be any time soon. Since fixed infrastructure

Table 13.1 Gas transport options for High Arctic stranded gas

Fuel type	Transport conditions	Energy density (BTU/ft^3)	Cost of transport (relative)
Methane (STP)	Not transported at STP	1,000–1500[1]	Not transported at STP
Pipeline	Compressed, rate of transport adds time factor	Depends on compression and velocity	Pipeline and pumping infrastructure costs high
CNG	Compressed	235,000 @ 205 Atm	Low
LNG	Compressed and Cooled 25 kPa −162 °C	563,000	High
NGH	−20 C 1 Atm	141,000–188,000	Competitive with LNG
Gas to liquids	Liquid ambient P–T	800,000–900,000 (est.)	D1, D2
Diesel	Liquid ambient P–T	939,000	D1
Gasoline (Petrol)	Liquid ambient P–T	805,000	D1

Diesel and Petrol are shown for reference. All figures based on Wikipedia, arbitrarily rounded to nearest 1,000. *D1* cost depends on the mode of transport (i.e., tanker, pipeline), *D2* cost depends on composition (i.e., percentage of higher density hydrocarbons in addition to methane). 1—Methane alone, usually some higher density hydrocarbons are present in natural gas, yielding a higher energy density. CNG and LNG are for methane. Cost of transport: Pipeline costs are very dependent on distance whereas ship costs less directly so. CNG depends on tank pressure; higher-pressure tanks cost more initially

to bring gas from NGH play zones may not be developed for some time, a variety of transport options may need to be assessed (Table 13.1). Mobile transport of gas produced from Arctic oceanic NGH will probably be necessary, at least at first.

There are two critical factors in the mobile transport of the gas, energy density of a particular volume of the material and the relative cost of converting the gas from its field production character to its transported form and the cost of reconversion, where this is necessary, at landing. Unlike petroleum, gas can be transported in a number of physical forms, each of which has different characteristics.

Stranded gas deposits, even including conventional gas supergiant finds in the high Arctic, will have the same character, once the gas has been extracted from its reservoir. Mobile transport systems will have to be based upon remote conversion of produced gas to another form for transport from the High Arctic (Conser 2007; Abdel-Kreem et al. 2009). Each transport method will have different costs but will likely have the same issues of distance from a shore landing and gas handling issues. Because of its extremely low environmental risk and high potential payoff, it may be that technology developed to commercially develop Arctic NGH in the near term could have the additional effect of promoting the development of gas transport infrastructure that could also be used for conventional hydrocarbon production in the High Arctic as exploration moves northward.

The cost of each mobile transport method has to be assessed in relation to the cost of conversion of the gas. In principle, gas can be transported more efficiently when the energy density achievable is higher. In practice, however, the cost of conversion of the gas to a higher density form may be a major component of the overall cost; the cost of the actual transport being about the same, once the capital cost for each method is normalized.

In addition to altering the energy density of stranded gas as a means for transporting gas from remote areas such as the High Arctic, the worldwide low wellhead cost of gas could be monetized in competition with more expensive petroleum. For instance, conversion of gas to liquid fuel would allow the value of gas to closely compete at retail with higher price oils. The market for gas is also increasing. Gas fired power stations have almost completely replaced oil fired stations in California, for instance, and gas is well on the way to replacing much of the one-time nuclear power generation capacity in Japan. In addition, gas is being used as a fuel in vehicles, particularly heavy vehicles and buses.

Compressed Natural Gas (CNG): The pressure at which the gas is transported is important. The higher the pressure, the more gas can be transported in a single shipload. Higher-pressure tanks, however, are more expensive than lower pressure tanks. Converting inlet gas to CNG involves simply the cost of compression, which has a calculable energy cost. If the inlet gas is already at pressure, the compression cost may not be extreme in the first instance. However, capturing the thermal energy produced during compression and using energy recovery electricity generation similar to that which is now dramatically lowering the cost of reverse osmosis seawater desalination, in which the high pressure product and reject water streams are used to generate electricity as part of lowering their pressure (Stover 2008), much of the cost of this compression could be recovered. In an extreme cost model in which the inlet gas pressure is higher than the line pressure at the point of delivery, the process could actually be energy positive. To our knowledge this cost reduction technology has not yet been factored into CNG transport options.

We view the factor of energy density as a function of cost of conversion as being relatively low because no new equipment development for loading or landing CNG is required. Special CNG ships would require design and manufacture. Container-based transport using multiple element gas containers offer promise of lowering startup costs has been proposed (TOG 2012), but may involve an additional cost of refrigeration.

Liquefied Natural Gas (LNG): The energy density of LNG is the present benchmark by which the other forms of gas transport can be compared. It is a well-understood form of natural gas that has been used for commercial transport for many decades. Production of LNG involves considerable energy but in addition, is usually done in very large installations on land that has pipeline access to large amounts of gas. The gas is compressed, with a very high production of heat requiring considerable refrigeration, so that it is condensed into a liquid at close to atmospheric pressure at approximately $-162\ °C$ ($-260\ °F$). LNG is transported as a bulk cargo in large tanks at a maximum pressure of about 25 kPa/3.6 psi); expensive pressurized tanks are not required. Vaporization of the gas, which is available to use in the engines, keeps the LNG cold. Design and implementation of a mobile, ship-borne LNG plant would involve a new initiative for miniaturizing

LNG manufacture technology and fitting it into a ship that might have its own LNG tanks. Transfer of LNG from one ship to another at sea introduces a level of risk that to our knowledge has not been properly evaluated.

Manufactured Gas Hydrate (MGH): It is possible to manufacture gas hydrate from gas produced from any source and transport it in a solid form (Kanda 2006). Pelletized transport has been suggested (Kanda 2006) for ease of manufacture and handling. The energy density of MGH is relatively low (Table 13.1) because gas is compressed within the MGH by about 164 x (Sloan 1998) and because the packing of the pellets leaves some space that is not filled with MGH. The porosity can be flooded with gas to help stabilize the MGH, but the pressure in the hold is envisaged as being only one atmosphere (Nogami and Nobutaka 2008). Although the costs of fabricating the MGH and dissociating to produce gas and water are substantial, these costs (including the value of the fresh water) may be economical enough to allow for MGH it be used commercially for transport. Mitsui Engineering and Shipbuilding of Japan, which is a major manufacturer of ships for transporting oil and natural gas, is taking the lead in assessing this method.

Gas to Liquids (GTL): The GTL process is an established technology that produces liquid fuels having diesel-like energy densities and naphtha from natural gas. The process was pioneered to produce liquid fuels from coal, a process that was perfected in Germany during WWII and is currently being used for the production of liquid fuel, most notably by Sassol in South Africa. Commercial GTL installations exist (Shell 2011; Siemens 2012) and shipboard GTL plant has been proposed by a number of companies, in particular 'Syntroleum'. Shipboard GTL plant would be an elegant solution for remote conversion of stranded gas (Hall et al. 2001).

There is no question that the energy density of GTL is the highest energy density obtainable for natural gas. Although the costs of large plant are well understood, commercially viable, miniaturized mobile GTL plant does not currently exist and would require a new technology development. GTL would also allow gas to be used to dilute heavy oils for greater ease of pipeline transport. To us, GTL is an attractive option because of its high energy density and ease of handling of a finished liquid fuel that will be directly usable upon landing without further processing or refining.

References

Abdel-Kreem M, Bassyouni M, Shereen M-S, Abdel Hamid, Abdel-Aal H (2009) Open Fuel Cells J 2:5–10. http://creativecommons.org/licenses/by-nc/3.0/

Conser RJ (ed) (2007) Topic paper # 20 deepwater. Working Document of the National Petroleum Council, submitted to the Secretary of the United States Department of Energy 18 July 2007, p 26

Hall KR, Bullin JA, Eubank PT, Akgerman A, Anthony RG (2001) Method for converting natural gas to liquid hydrocarbons. U.S. Patent 6,323,247 (issued 27 Nov 2001)

Kanda H (2006) Economic study on natural gas transportation with natural gas hydrate (NGH) pellets. Paper to accompany abstracts. 23rd World gas conference, Amsterdam, Holland, 5–9 June 2006, p 11. http://igu.dgc.dk/html/wgc2006/pdf/paper/add10399.pdf

References

Nogami T, Nobutaka O (2008) Development of natural gas ocean transportation chain by means of natural gas hydrate (NGH). In: Proceedings of the 6th international conference on gas hydrates (ICGH 2008), 6–10 July. Vancouver, British Columbia, Canada, p 9

Shell (2011) First cargo of Pearl GTL products ship from Qatar. http://www.shell.com/home/content/media/news_and_media_releases/2011/first_cargo_pearl_13062011.html

Siemens (2012) (dated as 2007) Process analytics in gas-to-liquid (GTL) plants. Case study. Siemens AG. www.siemens.com/processanalytics

Sloan ED (1998) Clathrate hydrate of natural gases, 2nd edn. Marcel Dekker Inc., Publishers, New York, p 705

Stover RL (2008) Energy recovery devices in desalination applications. In: Proceedings, international water association North American membrane research conference, Amherst, MA, 10–13 Aug 2008, p 8

TOG (2012) Trans Ocean Gas. http://www.transoceangas.com/

Chapter 14
Natural Gas as Fuel and Renewable Energy Aspects

Abstract Gas is the fuel of the future, especially in circumstances in which cutting CO_2 emissions to the atmosphere is an objective to slow the increase in greenhouse effect of the atmosphere. Natural gas could be envisaged as a bridge fuel from higher CO_2 producing combustibles to a non-CO_2 renewable energy situation. In addition, because NGH is generally in the process of crystallizing in the vicinity of existing deposits now, replacement NGH may form in relatively short periods of time within a deposit under production. This gives NGH the peculiar attribute of being a clean, renewable, combustible fuel.

Keywords Renewable • Fuel • Bridge fuel • Greenhouse • Climate change • CO_2 • NGH

Natural gas produced from NGH will be very clean burning and produces the least CO_2 per BTU of any combustible fuel. It could most reliably be considered to be a bridge fuel from the current mix of combustible fuels to a renewable energy economy that does not exhaust vast amounts of CO_2 into the atmosphere. Because of increasing energy demand, concerns for the climate that favor development of alternative energy sources that do not exhaust CO_2 into the atmosphere, and the limited nature of hydrocarbon fuels, major energy system changes are being implemented, albeit somewhat slowly. Renewable wind, water, geothermal, and solar energy that do not effuse CO_2 into the atmosphere are increasingly being seen as beneficial replacement for energy produced by combustion. This is a commendable objective but renewable resources are by their nature not full-time or wholly predictable. Therefore, base load, high energy on demand output power to back up alternative energy sources and fill any demand gaps is required in any conventional-unconventional power grid.

Natural gas is the best fuel solution for base load power in a dominantly renewable energy power spectrum for a number of reasons. Power plants can be relatively small and dispersed, with gas being delivered through low to medium pressure pipelines. Small gas fired power plants are off-the-shelf items that can be ordered and installed more rapidly than any other combustible fuel power station.

Once installed, they can be started or turned off quickly, as drops in renewable energy power varies. They can remain in a 'cold' state for long periods without extensive maintenance, and be brought up to full power much faster than either oil or especially coal, which has a very long startup time. In addition, less CO_2 per BTU is produced than from burning coal or oil and because natural gas is relatively pure and will burn at high temperatures, its exhaust is essentially CO_2 and water vapor with low ancillary pollution. 'Clean gas' is the standard base line against which all combustible fuels can be measured.

There is special aspect of NGH that renders it fundamentally different from other gas sources. Whereas conventional and the other unconventional gas resources were essentially produced long ago in geological time and trapped in geological traps that are unlikely to refill from their source beds, oceanic NGH formation and concentration, such as may be found in the continental margin zone (Figs 9.1 and 9.2) is dependent on recent or current adequate gas flux into suitable host beds residing within the GHSZ. The rate of new NGH growth will depend on the rate at which natural gas or groundwater carrying sufficient dissolved natural gas migrates into suitable hosts in the GHSZ.

Regeneration of NGH is possible because neither of the most likely potential NGH conversion techniques, heating and depressurization, significantly disturb the framework structure of the host sands, although some compaction may take place if sufficient back flooding to maintain reservoir pressures is not properly carried out. If the gas flux continues and the GHSZ remains in place, NGH will continue to form once conversion stimulation effects have come back into equilibrium with ambient conditions. In practice, there is evidence for NGH reformation in a zone of dissociation in the Mallik well in the Mackenzie Delta region of northern Canada, in the interval between periods of production. During the shut-in period from the end of the 2007 winter test to the beginning of the 2008 winter test, all the free gas associated with the dissociation of NGH during the 2007 winter test was absorbed to re-form NGH, which had the effect of increasing the MH saturation in the vicinity of the well by about 1–5 % (Kurihara et al. 2011). Thus, in a manner of speaking, NGH may be regarded as a renewable resource in which wells could be reentered for new phases of production on a scale of possibly decades. No other gas resource has this aspect of reformation, although secondary concentration of both gas and oil in high zones of conventional reservoir traps is known to take place (Selley 1998).

References

Kurihara M, Ouchi H, Sato A, Yamamoto K, Noguchi S, Narita J, Nagao N, Masuda Y (2011) Prediction of performance of methane hydrate production tests in the eastern Nankai Trough. In: Proceedings of the 7th international conference on gas hydrates (ICGH 2011), Edinburgh, Scotland, UK, 17–21 July 2011, p 16

Selley RC (1998) Elements of petroleum geology, 2nd edn. Academic Press, London, New York, Toronto, p 470

Index

A
Active margin, 36, 38
Alpha-Mendeleev ridge, 4
Amerasia basin, 3, 4, 5, 11, 12, 15, 16
Amundsen gulf, 12
Amundsen trough, 78
Anadyr river, 16

B
Banks Island, 2
Barents margin, 13–15, 82
    NW barents margin, 14
    West barents margin, 13
Barents sea, 1, 5, 14, 68, 82
Basin analysis, 55, 57
Bennett island, 15
Blake ridge, 50, 96
Bottom simulating reflector, 36
    identification, 63
    string-of-pearls, 39, 64

C
Canada, 12, 16, 25, 29, 48, 85, 91, 92, 110
Canada basin, 4
CAPEX, 70, 100
Cascadia, 36
China, 25, 28, 47
Chukchi microcontinent, 2
Chukchi sea, 2, 16
Cryosphere, 26, 28
Coalbed, 49
Coalbed methane, 48, 67
Compound NGH, 13, 20
Compressed natural gas, 105
Conventional gas deposits, 73

Conversion, 66
Cretaceous, 4, 15, 16

D
Depth, 1, 6, 9, 34, 77, 80, 93
Drilling, 28, 65, 66, 71

E
East Alaska, 2, 12
East Siberian sea, 4, 10, 12, 16
Ellesmere Island, 2
Energy density, 20, 49, 82, 105
Eurasian basin, 1, 3, 5, 6, 12, 64, 80, 83
Exploration, 3, 9, 19, 25, 26, 28, 29, 52, 55, 62, 65, 91

F
Flow assurance, 68
Fram strait, 2
Franz Joseph land, 13

G
Gakkel ridge, 5
Gas flux, 20, 63, 80, 110
Geologic trap, 22, 26–28, 33, 34, 39, 67, 70, 110
Geomechanical character, 23
Geothermal gradient, 11, 20, 27, 63, 83
GHSZ
    GHSZ thickness, 27, 41, 51, 61, 64, 78, 83, 86
Global energy assessment, 86
Greenland
    Greenland margin, 12

Growth dynamic, 21
Gulf of Mexico (GoM), 4, 51

H
Heat flow data, 63
Henrietta island, 2
Hyperpycnal, 44
Hyperpycnal flow, 44

I
India, 25, 29, 47, 100, 101
Indigrirka river, 16

J
Japan, 25, 29, 47, 52, 73, 100, 106
Joint industry project, 56, 100

K
Kansan, 42
Kara sea, 2, 13, 14, 100
Khatanga river, 15
Kolyma river, 16
Korea, 25, 29

L
Labrador current, 78
Labrador sea, 5, 51, 91
Laptev sea, 5, 13, 15, 16
Lena river, 15
Liquefied natural gas, 52, 103
Logging, 58, 71
Lomonosov ridge, 2, 30, 79
Louisiana, 42, 43

M
Mackenzie delta, 12, 27, 110
Manufactured gas hydrate, 103
M'Clure strait
   McClure straight, 2, 12
   Northwest passage, 15
Messoyakha, 49
Methane, 3, 9, 20, 26, 49, 91
Migration pathways, 93
Miocene, 13
MMBTU, 99, 100
Morris Jessup rise, 2, 5, 79
Mississippi river, 35, 42–44
Monterrey canyon, 42

N
Nankai, 10, 29, 47, 52, 56, 67, 72, 78, 82, 85, 86, 94
Nansen basin, 5
Natural gas hydrate (NGH)
   biogenic, 35, 65, 91, 96
   cage structures, 19
   characterization, 55
   commercialization, 99
      economic and political factors, 25, 99
      logistical factors, 68
   compound NGH, 13, 20, 26, 36
   depth, 1, 2, 57, 78, 82
   exploration, 10, 29, 51, 82
   growth dynamic, 22, 35
   NGH conversion, 29, 49, 66, 94–96
      chemical exchange, 94
      depressurization, 70
      dissolution, 22, 94
      geotechnical models, 67
      thermal stimulation, 26, 94
      thermodynamic models, 22, 66
   NGH deposits, 11, 26, 51
      feeder system, 34, 65
      fracture-filling, 49
      high grade, 16, 50
      low grade, 49, 88
      permeability, 21, 23, 95
      porosity, 38, 40
      vein-type, 28
   NGH-specific, 69, 72
   oceanic NGH, 2, 6, 21, 26, 29, 51, 77, 88
   permafrost NGH, 26, 28, 29, 67
   phase boundary, 22, 94
   pressure, 21, 70, 95, 96
   production, 23, 29, 49, 66–68, 70, 72, 74, 94, 96
   resource estimation, 62
   stability field, 26, 95
   temperature, 20–22, 28, 68
   thermogenic, 96
Nebraskan, 42
North American Arctic Islands, 2
North American plate, 5
North slope, 2, 27, 69, 85
Norway, 13, 48, 69, 92
Novaya zemyla trough, 2

O
Ob river, 15, 42
Olenka river, 15
Oligocence, 42

OPEX, 100
Outliers, 79, 80, 82, 87

P
Paleocene, 57
Passive margin, 36, 38
Pechora river, 14
Petroleum system, 3, 22, 28, 49, 51, 55, 56, 61, 65, 77, 80, 83, 85
Pleistocene, 9, 14–16
Pliocene, 13
Plio-Pleistocene, 12
Processing, 56, 61, 62, 70, 72, 73, 106
Prospect, 23, 29, 51, 55, 74

Q
Queen Elizabeth Islands, 12, 82

R
Reservoirs
   high-grade reservoirs, 40
Rhone canyon, 42
Russia, 5, 25, 48, 69, 92

S
Safety
   exploration risk factors, 3
   inherent geosafety, 93
   production risk factors, 3
Seabed, 71
Sea level, 10, 16, 17, 21, 22, 28, 50, 78
Sea of Okhotsk, 16
Secondary recovery, 25, 66, 93, 94
Sediment delivery systems, 15
Seismic analysis, 50, 51, 62
Seismic data, 49, 56–58, 61–65, 73
Sequestration of natural gas, 22
Severnaya zemlya, 13, 15
Shale gas, 26, 29, 48, 49, 99, 100
Shtokman field, 100
St. Anna trough, 13–15, 78, 82, 83

Stranded, 82, 103–106
Summer sea ice, 91
Svalbard, 1, 13, 14, 69, 82
Svalbard archipelago, 2

T
Taymyrskiy peninsula, 2, 13, 15
Tectonics, 3, 16
Tectono-sedimentary framework, 9, 10
Tight gas, 26, 48, 49
Transport, 10, 14, 16, 48, 74, 82, 104–106
Trough, 6, 10, 12, 14, 17, 51, 78–82, 86, 87, 101

U
U.S. Department of Energy (DOE), 47, 56
   NETL, 47
U.S. Geological Survey, 50, 56
U.S. Minerals Management Service, 56
UNCLOS III, 92
Unconventional gas, 26, 48, 49, 74, 100, 110
Undersea processing and completions, 72
United States, 29, 47, 48

V
Valuation, 56, 58
Vestnesa ridge, 14

W
Walker ridge, 10, 73, 74
West Siberian sea, 2
Wisconsin, 42
Wrangel abyssal plain, 4
Wrangle basins, 4

Y
Yana river, 15
Yenisey river, 15
Yermak plateau, 2, 5, 82

Printed by Publishers' Graphics LLC
LMO131026.15.13.23